NOT BY DESIGN

NOT BY DESIGN
THE ORIGIN OF THE UNIVERSE

VICTOR J. STENGER

PROMETHEUS BOOKS
Buffalo, New York

To Phyllis, Noelle, and Andy

91 90 89 88   4 3 2 1

Library of Congress Cataloging-in-Publication Data

Stenger, Victor J., 1935-
   Not by design: the origin of the universe/ Victor J. Stenger.
      p.   cm.
   Bibliography: p.
   ISBN 0-87975-451-6 : $22.95
   1. Physics—Philosophy. 2. Cosmology. 3. Astronomy. I. Title.
QC6.S812   1988
523.1—dc19
                            88-4162
                              CIP

Contents

Preface

Few questions pondered by humankind throughout the ages are as fundamental as that of the origin of the universe. Answers have been as
many and varied as the cultures whose unique world views have produced
them. The answers do not always require a god-like creator. The Maoris
of New Zealand imagined nine states of the void eventually leading
to darkness and light, and finally to earth and sky. Similarly the Samhyka
in India speak of space condensed out of the void. In one ancient Chinese
view, the elements were produced within a mist of chaos. Samoans
believed that their islands were fragments of a cosmic egg that exploded
from a landless, skyless void—a nebulous, purposeless space. To the
Greeks, the earth was born of the marriage of Chaos and Eros, before
the existence of the gods. In the Chinese Taoist picture, the universe
is the creative act of an inherent, natural and unknowable entity.

In our own culture, dominated as it is by Judeo-Christian thought,
most people normally take as evident that the universe began as the
result of some creative act, either on the part of an all-powerful deity
or some other transcendent entity. This traditional belief has its origins
in early superstitions, but has been buttressed over the ages by theological
and philosophical arguments. Thomas Aquinas believed he had proven
the existence of God logically, as the "first cause, uncaused." Even today
one hears this as a common argument for faith. Something must have
caused the universe; that something is God. Later philosophers, however,
have pointed out the error in Aquinas's logic: if a first cause, uncaused,
is possible, why must it be God? The first cause, uncaused, could just
as well be the universe itself.

Related to the causality argument is the argument from design, which
became popular after the scientific revolution in the seventeenth and eighteenth centuries. Then it appeared that the universe was a vast mechanism,

operating according to natural laws. The very order of the universe was seen to imply a *grand design*. The natural laws responsible for that design must have been laid down by some lawgiver. That lawgiver was God.

Modern science has even been called on to testify to the validity of the notion of creation. The galaxies have been discovered to recede from one another like fragments of an explosion that happened 10 to 20 billion years ago: the *Big Bang*. In a papal encyclical in 1951, Pope Pius XII took this as evidence for a Creator, saying: "Science has provided proof of the beginning of time. . . . Hence, creation took place in time. Therefore, there is a Creator; therefore, God exists."

In this book, this certainty is challenged. Nothing in our current scientific understanding of the universe, including the Big Bang, demands that it begin with a purposeful creation. The fact that the universe had a beginning, and is not static but has evolved to its present state, does not necessarily imply a grand design. On the contrary, developments in physics and cosmology during the 1980s, further developing the Big Bang picture, are leading us to a startling conclusion: The universe, including its current orderly state, can have appeared spontaneously out of nothingness.

The origin of the universe is really not a theological or philosophical question, any more than the motion of the planets or the operation of a steam engine. The universe is a real place composed of observable bodies—galaxies, stars, living beings, atoms—which are the proper objects of scientific study. The origin of this universe is a scientific issue, specifically a question of *physics*. Hence, this book is primarily concerned with physics and its closely related discipline, astronomy.

Unfortunately, even highly educated people in our society have little training in physics. So it is necessary for anyone trying to reach a broad audience with any physics topic to spend a great deal of time explaining things that would normally be covered at high school or early college levels. This book is no different. The reader already familiar with basic physics is asked to bear with me; I hope that he or she will find something interesting or unique in my approach to these questions.

At the same time, there is no reason for an author to underestimate the intelligence of readers, or their ability to follow a chain of reasoning. The first purpose of education is the development of critical and rational thought, and this can be accomplished in any course of study. Thus I have followed the lead of many successful popularizers of science, such as Steven Weinberg and Isaac Asimov, in assuming an educated reader not necessarily familiar with physics. Thus the material is not simply presented as a series of facts; rather, I have attempted to describe the whole process of thinking that leads to my conclusions, in a way that should be understandable to most readers.

This, necessarily, leads me into the history and methods of science, as well as many physics issues that may not, at first glance, seem to bear on the main topic. In fact, they all do. The conclusions drawn in this book are the product of a methodology that began to be developed thousands of years ago. Mundane matters, such as heat flow between bodies or the motion of a projectile, have direct bearing on the outcome. These must be discussed for completeness, but a comprehensive survey of physics is not necessary. Instead I focus on the elementary particles of the universe and the fundamental forces by means of which these particles interact.

I have avoided extensive use of mathematics in the text; it is possible to be logical without using algebra or calculus. However, it would be silly to write out the words, "energy equals mass times the speed of light squared," when the simple mathematical formula, $E = mc^2$, says the same thing so compactly. Similarly, I use scientific notation for very large and very small numbers. I trust the minimal use of mathematical symbols will not scare away readers who may not have had occasion to use them for many years. The mathematics here is simply a sort of shorthand.

I have not found it necessary to include a large amount of astronomy. This is not a book about the current universe, but about the universe of 10 or 20 billion years ago. Planetary surfaces, stellar atmospheres, and indeed the galaxies themselves developed long after the events I will describe. I believe that this emphasis—on physics rather than astronomy and on the issue of unplanned versus designed origin—makes this book unique among the many books that are continually appearing on the subject of cosmology. Nevertheless, I commend many of these other books to the reader; some are listed in the References.

I do not, however, recommend many of the other books on bookstore shelves, which assert the authority of modern physics in support of fashionable new world news or paradigms about humankind and nature. The revolution of twentieth-century physics, resulting in the overthrow of much of Newtonian classical physics, did not go quite as far as many writers claim. The four pillars of classical physics were *rationality, reductionism, causality,* and *determinism.* It is true that the notions of causality and determinism have been challenged, or even overthrown, by quantum mechanics. But the concepts of rationality and reductionism still stand stronger than ever, solidly reinforced by modern developments.

Rationality is nothing more than the use of clear thinking, critical analysis, and logic. Being logical simply means using words in the way that they are defined. I have no sympathy for the notion that there is room for the irrational or illogical in human communication. What can be the value of fuzzy thinking, or the use of words in ways inconsistent with their definition? I believe there must be a rational basis for all human endeavor; not just science but art, music, politics, religion, and any other social or cultural activity.

I will maintain in this book that the notion of reductionism also still stands, after almost a century of rebuilding the structure of physics. Many modern writers, including some scientists, have tried to make much of the fact that the macroscopic world exhibits a disequilibrium and irreversibility not apparent on the microscopic scale of the elementary objects that compose matter. They take this as evidence that the "whole is greater than the sum of the parts." This has become the philosophical basis for much of what is called *New Age thinking, holistic medicine,* and the like. Maybe these ideas will turn out to have some validity, but the evidence is simply not yet there to support the grandiose claims now being made.

Certainly a cathedral is, in some sense, more than a pile of stones, and we know of "laws of nature" that operate only for collections of particles. In fact, this book will begin with an extensive discussion of the notions of entropy and the Second Law of Thermodynamics, which describe the irreversibility of macroscopic phenomena and give time an arrow not evident on the microscopic scale. But we will see that physics still successfully describes all collective phenomena in terms of the behavior of the basic particles of matter. Irreversibility and time's arrow are nothing more than statistical effects resulting from the large number of fundamental particles present in macroscopic matter. Our understanding of these is totally reductionist.

Anytime something new is discovered in the laboratory concerning macroscopic bodies, the basis for the understanding of the new phenomena inevitably lies at the level of atoms or molecules. No case can be cited in which some new macroscopic principle had to be invoked to explain the results of a carefully conducted, reproducible experiment.

Simply put, no evidence has yet been found that the complex organization of what we see in the macroscopic world around us results from anything other than the operation of the rules that govern the elementary particles and the rules of chance. Further, intricate structure can develop from simple rules at the elementary level. I will show examples of this in the first chapter.

The damage wrought by quantum mechanics to the classical concepts of causality and determinism is critical to the major theme of this book— that the order of nature happened by chance. While quantum mechanics was truly revolutionary, no justification exists for many of the implications claimed by popular writers. For example, the way in which the observer enters into the definition of quantum states has led some to suggest that human consciousness is somehow connected with the fabric of the universe. This notion is used to justify belief in special powers of the mind, such as *extrasensory perception,* the existence of a "spirit world," or "astral planes," which have no basis either in theory or in observational fact. The mysteries associated with quantum mechanics often exist only in the writer's mind due to lack of comprehension of the subject. There is not a whit of scientific evidence to support these outlandish claims. The "observer" in quantum mechanics can be an unconscious machine just as well as a human being.

The classical mechanics of Isaac Newton still retains usefulness in a vast number of everyday applications. Nothing that has been so successful for so many centuries can be easy to demolish. Similarly, quantum mechanics has had better than a half-century of success in explaining and predicting a host of observational facts, so its basic tenets must have some element of fundamental truth. These tenets strongly imply that not everything that happens has an identifiable cause or is determined by what may have happened before. Atoms and nuclei change states in an unpredictable, undetermined fashion. Particles are sometimes here, sometimes there, with no force pushing them. Matter appears out of nothing and disappears into nothing. Only a few small steps are required to go from these statements to the spontaneous generation of the whole universe.

The science of the origin of the universe is still highly speculative and incomplete. My claim is not that the absence of a creative force has been, in any sense, "proven" but rather that what we now know about nature does not, as many people believe, require either a creation or a Creator. The simplest hypothesis that so far seems to explain the data is that the universe is an accident.

1

Order by Chance

We had the sky, up there, all speckled with stars, and we used to lay on our backs and look up at them, and discuss about whether they was made, or only just happened.

Mark Twain, *Huckleberry Finn*

In the days before city lights and air pollution obscured the view, the pageant of the heavens was the nightly experience of everyone. Early sky-watchers were so impressed by the orderly movement of the sun, moon and stars across the sky that they thought of heaven as a world more perfect than the one they walked upon. On earth a storm or earthquake might strike without warning and create great destruction. A man might be healthy one day and dead the next, of accident or disease. There was an unpredictability of events on earth not reflected in the sky.

As people became civilized, they learned to use the more certain movements of the heavenly bodies to predict events on earth, such as the change in seasons and the Nile flood. By observing the location along the horizon at which a certain star rose, they could determine the best time to hunt or to plant, and more dubiously, when to go to war. The people who could read the heavens were more powerful than their kings. Though their knowledge was riddled with superstition and meaningless ritual, these ancient priests gave mankind the first measure of control over the environment and began the development of science that over a few short millennia has changed the world.

If much of the world around us seems chaotic and unpredictable, there is also rhyme and reason on earth. A ball tossed in the air will follow a smooth parabolic path back to earth. Water runs downhill and not up. Living things reproduce their enormous complexity with

great precision. The regular movement of celestial bodies and the predictable phenomena that we see on earth give us confidence that we
live in a universe that is basically orderly.

As we observe our surroundings, bits of data representing patterns
of light or sound, smell, taste, or touch, are collected and organized
in our brains to form the basis of concepts about the world. We invent
notions of space, time, mass, and energy to help us classify this great
flow of data. These bits do not appear to be random; they occur in
patterns that our brains somehow are able to detect. Many of the patterns
we observe are simple, like the arc of the sun's path, but most are
intricate and complex.

This orderliness, and especially the intricate structure and enormous
variety of life, is often taken as evidence for a *grand design*—an intelligence
beyond our experience that is responsible for the patterns of the universe.
"How is it possible that all this could have happened by chance?" people
ask. This question often serves as a common justification for faith in
a supreme being. Even in modern science, where gaps in our knowledge
of the natural world are no longer filled in with theology, the lexicon
has a built-in prejudice: the universe is governed by a set of rules, or
laws of nature, that have always existed. Scientists speak of the Law
of Inertia or the Second Law of Thermodynamics as if some great
legislature in the sky once met and set down rules to govern the universe.
This prejudice is primarily the product of our cultural traditions and
is unjustified by the factual data that the universe presents to our senses.

The existence of order does not necessarily imply that it is the result
of design. If parts of the universe now exhibit a certain structure, and
if there are patterns that can be described by scientific principles, this
order still could have "just happened." Furthermore, it probably did.
Nothing currently known about the universe requires that its structure
was somehow imposed upon it from the outside. The joint efforts of
physicists and cosmologists are now converging on a picture of how
the patterns now observed could have come about as a natural process,
accidental and devoid of plan.

Let us begin by clarifying what is meant by *order*. Even the cards
dealt from a shuffled deck have an order or sequence once they are
laid out on the table; but what we generally mean by order is a perceived
pattern in otherwise random events. How do we decide that a pattern
exists? Usually we make an instinctive and highly subjective judgment

that what we see is too unlikely to have happened by chance. In principle, though, the pattern can simply be the observation of improbable but still possible occurrences. Whether a possible event is then observed to occur is a matter of chance. Highly likely events are observed more often, but highly unlikely events can also occur. Play enough poker hands and one day you will be dealt four aces or a royal straight flush.

Jackpots are won daily on the slot machines of Las Vegas and Atlantic City. Each by itself is an improbable event, but with thousands of people pulling thousands of levers hundreds of times a day, simple chance will produce a few big winners on a regular basis. In the state lotteries that have become popular in recent years, millions of dollars are regularly won by individuals. The odds against each winner are astronomical, yet the games are structured so someone has to win. So with the universe. Starting out with no rules and everything possible, why shouldn't a universe like ours have formed? And many others not at all like our own as well?

Let me illustrate how patterns are produced by chance. I have programmed my computer to draw a picture in which there is an equal chance that a dot or a blank space will occur at each point on the screen (fig. 1.1). Looking at the result as a whole, you see it for what it is—a random distribution of dots. But if you scan across the page you will find localized regions where intuitively unlikely wide gaps, or continuous dark bands, form some kind of visual pattern. For example, my eye detects a V-shaped pattern, with the vertex of the V about one quarter of the way down and about halfway across, one leg of the V continuing across to the right and the other diagonally down and to the right. You should be able to pick out other patterns; the human brain seems to be exceptionally suited for this sort of visual pattern rcognition. Do we just imagine these patterns? No, they are really there—put there not by plan but by chance.

Still it can be argued that only the simplest patterns, such as my computer-produced V, can occur by chance in any reasonable time. For example, Fred Hoyle and N. C. Wickramasinghe estimate that the probability that a single random trial would produce the enzymes needed for life to be created in the ancient oceans is $10^{-40,000}$ (Hoyle 1981, p. 24). They conclude that life could not have formed in the way proposed by current scientific consensus.

This calculation assumes complete randomness, which certainly did

Fig. 1.1 The human brain, designed as it is for pattern recognition, can imagine patterns in random data. Here a display of random points is shown as plotted on a computer screen. Each point was plotted with a fifty-fifty chance of being a dot or a blank, independent of any other point. Numerous patterns can be seen. For example, a V on its side is indicated on the upper right going almost halfway across the picture.

not exist in the earth's ocean at that time. Although a long time ago—about 4 billion years—this was still billions of years after the universe began. By that time gravity and electromagnetism existed, and the molecules in the ocean were bound by their rules. If the origin of the universe was accidental, this does not imply that such rules do not now exist, or did not exist at the time life first formed. Rather, this hypothesis implies only that no rules existed at the beginning of time; instead, they came into being spontaneously a tiny fraction of a second afterward. Once a few simple rules exist, remarkably complex structures can readily form spontaneously in surprisingly few trials and periods of time well within the finite age of the universe.

The advent of computers has made possible studies of the spontaneous generation of structure in fields as diverse as biology and astronomy. Let me give an example of one pattern that occurs frequently in widely different contexts: the spiral. Spiral patterns exist in living systems such as slime molds, the cross sections of nerve axons, the eyes of fireflies, and seashells, to name just a few. The stars arrange themselves in spiral patterns in most galaxies. Some chemical reactions, called Belousov-Zhabotinskii reactions, have been observed to spread in a spiral fashion.

The spiral behavior of Belousov-Zhabotinskii reactions was first observed in Russia in 1958. More recently, two astronomers trying to understand galactic structure, Barry Madore and Wendy Freedman, have performed computer simulations showing how spiral patterns follow naturally from a few simple rules (Madore and Freedman, 1987).

Anyone with a home computer and some programming ability can duplicate the Madore-Freedman results, and I will explain them through a simple computer experiment. Consider your display screen area to be divided up into small cells. Each cell can exist in one of three states: *active* (*A*), *receptive* (*R*), or *quiescent* (*Q*). Active (*A*) cells have the ability to excite any neighboring receptive (*R*) cells, but not quiescent (*Q*) cells, into the active state.

Each *A* cell will be displayed as a point of light. Start out with all the cells in the *R* state, except for a slight "perturbation" that breaks the overall symmetry of the screen: set a line of about five cells in the center to *A* and an adjacent line to *Q*. Then start a time sequence in which any cells active for one cycle are turned quiescent, and any cells quiescent for two cycles are turned receptive. Updating the display

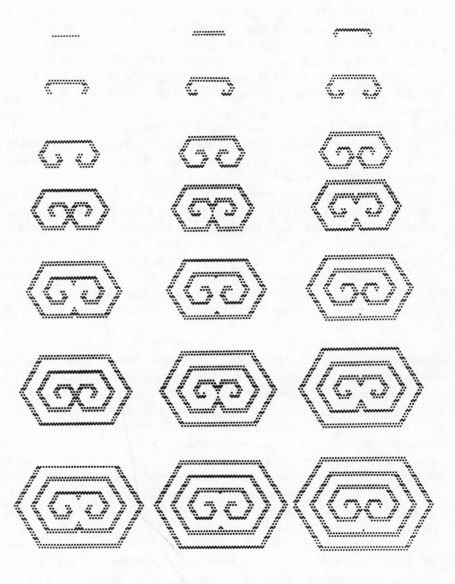

Fig. 1.2. The simulation of a "self-organizing structure," the Belousov-Zhabotinskii reaction, by a simple computer algorithm as described by Madore and Freedman (1987). Starting at the top left with a single line, the ends of the line are seen to curl up into spiral waves that propagate outward.

each cycle, you will observe spiral waves emerging from the ends of the line, curling around these ends, and propagating outward (fig. 1.2). This is precisely what is observed in the laboratory for Belousov-Zhabotin-skii reactions.

These results are not obtained at random, since we have introduced the rules that govern the behavior of the cells in the three states. Thus the resulting patterns are not accidental. Although rules are used, they are very simple ones that do not contain the mathematical equation for a spiral. In fact, using the same rules but with a slight change in the perturbation gives quite different results. For example, instead of starting with two lines of active and quiescent cells, cause a single cell to cycle between active and quiescent states. In this case, you will generate a series of circular waves radiating outward from that cell. Do this at several points, and you will see waves radiating from each point, merging with one another to produce many different shapes and patterns. The pattern produced depends less on the rules than on the perturbation, which could just as well happen by chance.

You can easily put more chance into your demonstration by randomizing the perturbations that trigger the pattern formation process. If you are really clever, you can even randomize the rules, making the whole process random. In that case, you would be generating your own *model accidental universes*, with patterns developing by whatever chance rules your random number generator happened to pick.

Of course we expect many different and unpredictable patterns to occur as we randomly change the rules. Most will be rather dull and you will probably quickly lose interest in just "tossing dice." Like the bored Creator in some creation myths, you will soon start to invent rules that yield beautiful and interesting patterns. This game of simulating the generation of patterns and structure on the computer is now widely played, and many serious papers have appeared in the scientific literature. There are numerous applications in understanding not only the formation of structure, but also the evolution of many particle systems.

Another point is illustrated by computer experiments: The generated pattern is often simple until a random break occurs, which then leads to more complex and unexpected patterns. This is called *broken symmetry*. In nature, this can happen by natural fluctuations and need not be induced from the outside. This association of structure with *broken symmetry* has great importance for our understanding of how the natural order of the universe came about.

Computer games aside, how could the structure and laws of nature, seemingly so complex and yet so elegant and universal, possibly have been an accident? As illustrated above, the rules need not be as complex as they appear, and complex structure can be obtained with simple rules. The late physicist Richard Feynman likened nature to a chess game in which incredibly complex situations arise from a few simple rules.

Before we can find the simple rules that describe complex phenomena, we need to find the simple rules that describe simple phenomena. Let us then focus on the simplest type of orderly behavior we observe in the world—that of a single body.

The motion of bodies is the first thing a student of physics learns about, because it is easy to understand and forms the basis for discussing the behavior of the collections of bodies that make up the world. The straight-line motion of a coin dropped from the teacher's hand, the parabolic path of a dart fired in the air at some angle with the horizontal, and the elliptical paths of the celestial bodies are all examples of regular motions, which the equations of Newtonian mechanics successfully describe.

Encouraged by this success, we jump to the unjustified conclusion that these equations are in fact "laws" of nature, and that the bodies are somehow required to follow the paths predicted by the equations. If there were no laws in operation, the bodies would be expected to dart this way and that, with any direction possible at any time. The fact that they do not leads us to conclude that a "law" is in operation.

Suppose, however, that there are no laws in operation. Given a large number of bodies moving around randomly, if one waits long enough, it can happen that a few bodies will be found to move along a smooth path for a short time. The random motion of a body in a smooth line or curve is not impossible, just improbable.

The reason people assume that there is some directing power, intelligence, or law behind perceived orderly phenomena such as the smooth motion of bodies is that such order seems much too unlikely to have occurred by chance. There are many phenomena of this nature which, while not strictly impossible, are so unlikely that we can safely consider them impossible for all practical purposes. The air in a room will not all rush out, leaving behind a vacuum, when a door is opened;

yet it could happen, killing everyone inside. Each air molecule simply needs to be heading in the direction of the door when it is opened. Or, suppose a glass falls off a table and smashes into a thousand pieces. All the pieces can happen to be moving in just the right direction to reassemble into a glass and fly back upon the table—admittedly unlikely, but not impossible.

It is even possible for an old person to become young again, or a dead person to be brought back to life, with just the right movement of molecules. This would be regarded as a *miracle*, yet such incredible events are not, strictly speaking, impossible. Observing a miracle of this nature, we would be incapable of proving it as such, and not simply a result of blind chance. We could calculate the probability of the event occurring and get a number so low as to make the event unlikely in many times the age of the universe. But such calculations after the fact cannot be used in any impossibility proof. If they could, we all would prove ourselves out of existence, since the existence of any given individual is so unlikely. Nevertheless, none of us should count on spontaneously growing younger or coming back to life once we are dead, despite the fact that such a miracle is not impossible.

Let us move to more mundane examples of everyday life, which constitute our main body of knowledge about the world. First, consider what occurs when a hot body is placed in contact with a cold one. According to experience, heat will always flow from the hotter to the colder body until the temperatures equalize. This indeed has happened in countless observations throughout history, with no reported case to the contrary. Yet, a cold body contains heat energy in the motion of its particles and there is nothing *in principle* preventing the colder body from transferring that energy in the other direction, to the hotter body, raising the latter's temperature further while lowering its own. If this were to happen, it would be like heating or cooling a room without the expenditure of energy. This again is a highly unlikely but not impossible event.

In physics this apparent unidirectional behavior of everyday phenomena is called the *Second Law of Thermodynamics*. In addition to describing the fact that we do not observe heat flow from colder to hotter bodies—unless there is an expenditure of energy, as with a refrigerator or air conditioner—the Second Law explains why no one seems to be able to build a perpetual motion machine. Energy must always be expended in the engines we build to lift and move objects; no engine has perfect efficiency, converting all its energy to useful work.

These statements of the Second Law are often expressed in terms of a quantity called *entropy*. We will have much occasion in this book to discuss entropy. In terms of entropy, the Second Law of Thermodynamics states that *macroscopic events happen in such a direction as to increase, or keep constant, the total entropy of a physical system.* For our purposes, entropy can be simply translated into *disorder*. If one part of a system is to be made more orderly, this must be accompanied by an increase in entropy, or disorder, elsewhere. This increase will at least compensate, and in fact usually overcompensates, for the localized decrease in entropy. That is, local order can be produced, but only at the expense of an equal or larger increase in the total disorder of the universe as a whole. Usually this happens with the expenditure of energy.

A common misunderstanding about the Second Law of Thermodynamics causes it to be frequently misused in arguments supporting the idea of special creation. It is often said, for example, that Darwinian evolution violates the Second Law. In fact, it does not. The production of orderly structures, such as living beings, on earth results when ordered energy from the sun is radiated back to the universe in a more disorderly state, so that the total entropy of the universe increases.

Entropy, as disorder, is a measure of our lack of information about a physical system, although what that has to do with heat engines and refrigerators may seem a bit obscure. Actually the connection is not difficult to see. We regard a system as orderly when we know a lot about it, and disorderly when we know little. For example, a heat engine takes energy from its combustion chamber in an ordered form, uses some to do work, and exhausts what remains back into the environment in a less orderly form. So the total entropy of the engine and its surroundings increases, with a net increase in the total amount of disordered energy in the universe. Similarly, when heat flows from a hot to a cold body, the net entropy increases as the system becomes more disorganized; the order of the hotter body increases, but the colder body gets even more disorderly, with the net effect that the system as a whole is more disorganized than it was originally.

Technically, the entropy of a physical system can be complicated to calculate. However there are many cases where the entropy of one system is obviously greater or less than another. For example, suppose I have two chambers, one twice the volume of the other and each containing the same number of particles. The entropy of the particles in the smaller chamber is lower, because I have more information about the particles: they can be more accurately located than the particles in the larger chamber.

When an automobile tire is inflated it has a higher pressure than the outside air, resulting from the fact that it contains more particles than an equivalent volume of air outside. Now suppose the tire is punctured. Air will escape into the environment until the pressure in the tire equals that of the air outside. In the process, the entropy or disorder of the total system (that is, tire plus outside air) increases as the information we have about the air molecules decreases. Originally we know that a certain number were in the tire; afterward they are scattered throughout the atmosphere. We fix the puncture, but cannot wait long enough for sufficient particles of outside air to be moving in the right direction to go through the valve and reinflate the tire, although this is in principle possible. So we pump up the tire, an action that requires the expenditure of energy. In doing so we decrease the entropy of the tire-atmosphere system, but we must now include the pump and its energy source as part of the total system. The pump used energy that was produced somewhere, with an accompanying increase in entropy.

This is a characteristic result. A physical event can happen in the direction of decreased entropy, or increased order, but the numbers of particles involved in the phenomena of ordinary experience are so large that the likelihood of such an event happening spontaneously is, for all practical purposes, zero. To make it happen in a reasonable time, some agent outside the system must act, providing the force and energy needed to get the particles back to a more orderly state. When we build a building, we use energy to drive the engines, or the construction workers' muscles, to move the scattered bricks and pieces of wood into the pattern of the blueprint. If we could wait long enough, those pieces would assemble into the building with no help from engines or muscles. But we are impatient and use the energy instead. Ultimately this energy came from the sun in the form of highly ordered directional radiation. In using this energy to order environmental and biological systems, the earth returns energy to the solar system in the form of more disorderly lower energy radiation, emitted in all directions.

Beyond the obvious practical implications of the Second Law lies an even deeper philosophical statement. Everyone has seen movie clips run backward, with hilarious results: a diver flies from the pool back to the board; a broken window reassembles as the ball returns to the child's bat. Consider those spectacular films of a building being demolished with a few well-placed blasts. We could run one of those films backward and see the building apparently spontaneously built from broken pieces of concrete and wood. We would laugh at the ridiculousness of that. Or take our example of the air in a punctured automobile tire. We would not find it as funny as Charlie Chaplin walking backward, but if we ran a film of the tire backward, we would see air from the outside reinflating the tire without the action of a pump.

All these events, which we regard as impossible in normal experience, are artifacts of our running the film in the wrong direction. But suppose the tire requires only two air molecules to inflate it. We watch a movie of the tire and see two molecules from the outside flow through the valve into the tire. Can we conclude that the film is running backward? How do we know that the film is not really being run forward and it just happened that two molecules found their way into the tire? We cannot know.

In other words, *we cannot tell the direction of time from watching a motion picture of the movements of a small number of particles.* Only if the number of particles involved is large, like the 10^{24} or so in the everyday objects of our experience, can we distinguish one time direction from another. What Sir Arthur Eddington called the *arrow of time* is determined by the fact that events requiring large numbers of particles occur in the most likely direction, that of increased entropy. From our experience, we assume that time flows in the direction in which glasses break, tires deflate, and people age. Looking at it another way, the future is by definition that direction of time in which events are more unpredictable, where phenomena happen in the sequence for which we have less information.

In a hypothetical universe with a small number of particles, the flow of time in one direction would be (except possibly for small effects at subatomic levels, which need not concern us here) indistinguishable from flow in the other. So whether or not time has one particular direction seems to depend on the number of particles involved. Where do we draw the line? At 100 particles? 1000? 10^{20}? It is evidently arbitrary to say that time has a preferred direction when the number of particles is large, but not when the number is small. What is true for two particles should be true for 10^{24}. If time has no arrow for two particles, it has no fundamental arrow, simply one of convention, for 10^{24} particles.

What does this say about our need for an external agent to order a system, to reduce its entropy? When the film is run one way we need an agent. When it is run the other way, we do not. But why should it make a difference which way we run the film? How can the need for an agent depend on the arbitrary choice of time's arrow?

In the examples discussed so far, we can usually see the agent who provides the energy to order the system. We can watch construction workers build the buildings, or the garage attendant pump up the tire. We can even see many examples where an outside agent, such as a terrorist bomb, creates disorder instead of order. The question is not whether agents of order or disorder exist, but whether there is need for one when you see none. In fact, there is no need when there is no independent evidence. The argument that organization requires an organizer is invalid. *Order can occur by accident.* There are innumerable examples of this in everyday life, from the pattern of a snowflake to the chance encounter between a sperm and an egg that produced each of us.

This brings us back to the universe as a whole. A film of the energy of the universe would show an infinitesimal point of light exploding into an expanding shell of radiation. This is called the *Big Bang*. The radiation expands and cools, and after a few billion years or so we see a vast space uniformly filled with its very cold remnant, along with a leftover debris of matter clumped into clusters of galaxies of stars shining by the light of nuclear reactions at their cores. As we watch, the universe continues to expand, with the galaxies moving farther apart and the stars gradually burning out as their fuel is exhausted. Most die quietly, but others explode in supernovae, creating the heavier elements and spreading their matter around. New stars occasionally form and, when there are sufficient heavy elements, earthlike planets can also form.

When we run the film "forward," that is in the direction of our conventional choice of the arrow of time, we are presented with no problem as to the need for an invisible agent. The total entropy of the universe steadily increases or remains constant, *by definition,* even as little packets of order such as galaxies, stars, and planets are produced. This is possible because, as the universe expands, the maximum allowable entropy also increases. The Second Law of Thermodynamics is not violated when tiny ordered structures such as the earth appear; there is plenty of room for an increase in entropy of the rest of the universe to compensate for this relatively small entropy decrease. However, near the beginning of the film more profound issues arise: The universe appears in a nothingness, not only empty of matter and radiation, but equally without space or time. Can that have happened without an agent, a *Creator*?

Most of our ideas about the physical universe are expressed in terms of concepts such as space, time, and mass, which can only have meaning after the universe has reached a certain minimum size—the *Planck Length,* equal to 10^{-33} centimeter. This was reached when the universe was 10^{-43} second old, the *Planck Time.* We need not worry about what happened before the Planck Time: There was no before because there was no time. Time is a human concept, based on our limited experience. Our experience has so far not included distances and times so small that no clock to measure time as we know it can exist.

For time to have any meaning, we need a way to measure it— there must be a clock. Time is what is measured on a clock, and if there is no clock there can be no time. Before the Planck Time the universe was so small that even the concepts that have been developed in this century to explain the interiors of atoms and the interiors of the nuclei of atoms will not work because they cannot be defined in a logically consistent way.

Today we know that in the quantum world of atoms and subatomic particles a box containing no matter or energy is still not empty. Particles are created and destroyed on tiny time scales—typically less than 10^{-20} second—while the total energy remains zero on the average. If particles can be created on this time scale, then they certainly can be created on the scale of the Planck Time. The universe could have happened as a quantum fluctuation in nothingness. Just after the Planck Time, when we can safely use the concepts of space and time, the universe was a sphere of 10^{-33} centimeter, empty of radiation or matter as we now know them; today's particles and forces did not exist. Then, as the sphere expanded, order was spontaneously created as *elementary particles—*

quarks, electrons, neutrinos—and the physical laws they obey appeared out of nothing. None of this violates any known principle of physics. In fact, these principles themselves came into existence as part of the same ordering process.

Our universe with its particular structure was, by itself, a very unlikely occurrence. However, consider that not just one universe was created but many, each of them a little empty domain. One of these domains expanded and cooled, developing into the somewhat orderly universe we experience today, with one dimension of time, three dimensions of space, matter, and the rules that matter now appears to obey. Other universes may exist with different structures and rules, even different dimensions.

In many of the other domains, there may have been insufficient order to allow forms of organized matter to develop. Our domain was able to produce life on at least one planet. Very unlikely, yes—more unlikely than winning a state lottery or a slot machine jackpot. But there were countless domains, and the fact of our organized existence implies that we are in one that contains the patterns necessary for life. There may be many other universes without living beings able to reflect upon their existence. This idea—that our universe looks as it does because the fact of our own existence selects out a particular set of properties— is called the *anthropic principle*. Precisely those properties lead to life and living beings as we know them; in their absence we would not be here to reflect upon them.

So the universe could have happened by accident. But we cannot stop here. We want to know *how* it all happened, even if it was by accident. Current studies in elementary particle physics and cosmology have begun to develop provisional ideas of how the universe may have happened to take the form it exhibits to our eyes and our scientific instruments. The road of knowledge that has reached this point began about 2,500 years ago, hit a detour for about 1,500 years, and then finally began to move ahead again about 500 years ago. This is the path of *physics*: the scientific study of the fundamental structure of nature. We will now travel this path and see how far we have come from the days of ancient peoples who needed supernatural forces behind every tree and rock to explain the mysteries of the world around them.

2

The New Priesthood

The universal order, symbolized henceforth by the law of gravitation, takes on a clear and positive meaning. This order is accessible to the mind, it is not preestablished mysteriously, it is the most evident of all facts. From it follows that the sole reality that can be accessible to our means of knowledge, matter, nature, appears to us as a tissue of properties, precisely ordered, and of which the connection can be expressed in terms of mathematics.

Léon Bloch, *La Philosophie de Newton*

A thousand years, or even a million, is a short time when compared with the billions of years that have elapsed since space and time began. The human race is still in its childhood; recorded history bears about the same relationship to the age of the earth that a second does to a year. Children are completely self-centered creatures, aware of little beyond their immediate experience. The baby in the cradle is the center of its own universe, and its parents and everything else circles around it. So it was with the ancient observers of the sky. If heaven was more perfect than earth, still it revolved around the earth. Most ancient thinkers placed the earth at the center of the universe with the sun, stars, and planets revolving about it. One exception, according to Archimedes, was Aristarchus of Samos (3rd c. B.C.E.), who suggested that the earth revolved about the sun. But this idea was not accepted by the most influential thinkers of ancient Greece, Plato and Aristotle, and so did not become a component of collective knowledge for another millennium.

Plato (c. 429–347 B.C.E.) believed that the celestial objects moved in circles about an earth that was fixed and immovable. He was aware that certain objects, called *planets,* do not follow a straight path across the sky like the sun, moon, and stars, but rather wander about. To Plato this wandering motion presented no great problem; he argued that the true paths of all heavenly bodies were required to be circles

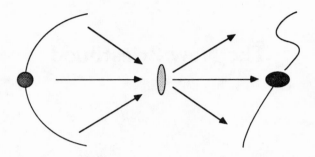

Fig. 2.1. In the Platonic view, the underlying perfect reality is distorted by our senses into the imperfect images we see. As illustrated here, the "real" orbit of a planet is a perfect circle that is distorted to the observed S-shaped path by the lens of the eye.

since their natural motion had to be described by that most perfect of geometrical figures. What we see with our eyes, said Plato, is a distortion of the true, ideal, underlying reality (fig. 2.1).

The ancient Greeks gave birth to the idea that the order of the universe is fundamentally a geometric order. This is not too different from the view of modern physics, which describes many phenomena in terms of geometrical symmetries and the way in which those symmetries are broken. For example, a sphere is symmetrical about any axis drawn through its center. For many purposes we can treat the earth as a sphere. Yet, the earth is really an oblate spheroid, fatter at the equator than the poles, still fairly symmetrical about its rotation axis, but with the symmetry about other axes broken. Nevertheless we can make scientific calculations assuming the earth is a sphere as a first approximation, and then compute the effects of oblateness as corrections.

Plato's view demonstrates a vital difference from what has become recognized as the scientific point of view toward "truth" or "reality." Present-day scientists are normally not allowed by their peers to ignore data presented to their senses when these data do not fit in with their preconceived ideas. Plato, and indeed much of humanity then and now, believed in an ultimate reality beyond the senses. By contrast, science has come to view physical reality as nothing more than what is accessible to the senses, either directly or aided by instruments.

Even the severest critics of science cannot deny its enormous success in understanding and controlling our environment, and predicting future events. Plato's rationalization that planetary orbits must really be circular appeals to our sense of symmetry, but it is not very useful. Assuming that the planets move in circles, when they do not, does not help one in trying to predict the motion of planets. Suppose an ancient astrologer told his king that he should begin a war Friday at 10 P.M. because that was when his calculations told him that Mars, the god of war, would rise above the horizon. If he used a circle to calculate Mars's path, and it did not rise as predicted, and the king lost the battle, he would likely be beheaded.

Claudius Ptolemy, who lived in Alexandria around the year 140 C.E., showed how to predict planetary motion with great accuracy. The Arabs, who preserved his work for posterity, called it the *Almagest*, meaning "the greatest of books." Ptolemy based his very elaborate model of the heavens on earlier ideas of Apollonius (3rd c. B.C.E.) and Hipparchus (2nd c. B.C.E.). He pictured the planets as moving in circles not centered on the earth, but themselves moving in circles around the earth. These circles upon circles were called *epicycles*, and as many as nineteen were needed in some cases to describe the motion of a single planet. Despite its complexity, the model was successful in accurately describing and, more importantly, predicting planetary motion. Epicycles explained how a planet viewed from the earth moves across the sky and then doubles back before turning once more to continue on its original path (fig. 2.2).

A much simpler model was proposed by Nicolaus Copernicus, born in Torun, Poland, in 1473. Copernicus returned to the idea of Aristarchus, that the earth was just a planet and, like the other planets, revolved in a circle around the sun (fig. 2.3).

At this point it is worth examining a fundamental issue concerning the "truth" of scientific theories. If two theories explain the same facts equally well, which one do we accept as truth? If both work, if both are useful, then we can feel free to use either. We certainly do not hesitate to use a geocentric point of view in describing earthly phenomena. For example, celestial navigation utilizes geocentric coordinates—it's the simplest way to do it. It would make no sense to express the motion of a ship or aircraft in a sun-centered coordinate system. But in the case of the planets, Copernicus's system is simpler than Ptolemy's, so it is the one we use in describing planetary motion.

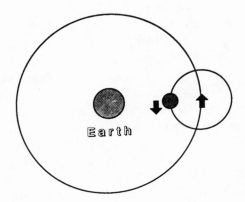

Fig. 2.2. In the Ptolemaic system, the earth is at the center. Planets travel in circles or "epicycles" whose centers travel in circles around the earth, thus explaining the planet's "wandering" path across the sky.

This idea—that the simplest theory rules—is called *Occam's Razor*, after the rebellious fourteenth-century Franciscan friar, William of Occam, who applied it to challenge both the Church and philosophers alike. Occam is quoted as saying: "Entities are not to be multiplied beyond need." In other words, we should not introduce more than the minimum number of concepts required to explain something. In modern science we use Occam's razor to cut away excess fat from a theory, always reducing and simplifying it to the barest essentials. When two theories that explain the same facts collide, the simpler of the two is provisionally accepted as the closer to the truth.

Copernicus did not publicly proclaim that the earth is at the center of the universe. In the preface of Copernicus's book, *De revolutionibus orbium coelestium*, published just before his death in 1543, the publisher Osiander carefully hedged by saying that Copernicus was only proposing a calculational tool: "The master's . . . hypotheses are not necessarily true; they need not even be probable. It is completely sufficient if they lead to a computation that is in accord with the astronomical observations." This kind of qualification is implicit in all scientific work today, and at the time proved wise. In 1600 Giordano Bruno was burned at the stake for preaching that the sun was the center of the universe and that the other planets had life upon them.

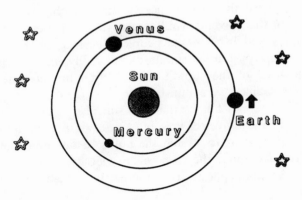

Fig. 2.3. In the Copernican system, the sun is at the center. Planets travel in circles around the sun.

Shortly thereafter, Galileo Galilei (1564–1642) of Florence made the first astronomical observations with the newly invented telescope. Observing spots on the sun and craters on the moon, he saw that these were not the perfect spheres that heavenly bodies were supposed to be. Seeing the phases of Venus and four moons orbiting Jupiter, he recognized that these facts would be difficult to explain in any earth-centered theory and concluded that Copernicus's picture was not just a mathematical model, but a better—or, by Occam's Razor, simpler—description of the way things really are.

In 1632 Galileo presented the Copernican view in a book named, in the wordy style of the day, *Dialogue . . . Where in Meetings of Four Days, Are Discussed the Two Chief Systems of the World, Ptolemaic and Copernican, Indeterminately Proposing the Philosophical and Natural Arguments, as Well on One Side as on the Other* (commonly known as the *Dialogue on the Two World Systems*). Actually the dialogue was not "as well on one side as on the other"; the Ptolemaic view is ineffectively argued by a character named Simplicio who is completely devastated by Copernicans named Sagredo and Salviati. To make matters worse, Galileo has Simplicio use words taken almost exactly from a statement by Pope Urban VIII, that the tides cannot be taken as proof of the motion of the earth "without limiting God's omniscience."

This did not help Galileo's case when he was tried and forced by the Church to recant his teachings. On the heels of the Reformation, the Church worried that acceptance of the idea that the earth is not at the center of the universe would cause further losses in its influence over the minds and lives of the people. Recently, after 400 years—a long time perhaps by human standards but really short considering the cosmic time scale of the subject—the Church has admitted that the actions against Galileo were a mistake.

Today's telescopes tell us that the sun also does not occupy any special place in the universe; it is but one star in a countless number in a cosmos so vast as to stretch the limits of our imagination. What was revolutionary about the Copernican view was its recognition that the heavens are not separate from the earth; instead, the earth is part of the heavens.

It is difficult for us to appreciate today what a profound and difficult concept this was for inhabitants of seventeenth-century Europe. The Judeo-Christian teaching put humankind at the center of all things in a universe created by God as humanity's domain. Suddenly all this was swept away.

These events happened in Renaissance Italy and were a product of that great awakening of the human spirit. Unfortunately, the trial of Galileo put a chill on further intellectual and artistic progress in Italy, and the foci of these activities moved to France and Protestant England and Holland. In England, especially, there had developed an amazing freedom of thought. This intellectual climate enabled Isaac Newton (1642–1727) to build upon the work of Galileo and initiate the developments that led to the scientific and technological world that we know today.

Galileo had deeply grasped a concept that was recognized by the ancient Greeks but lay dormant for some 1,500 years before being rediscovered in the late Middle Ages: *we must trust and believe what we see with our eyes, above any authority*. If what we see disagrees with what was written by some ancient revered master, then it is the master who is wrong and not our eyes. This principle, that observation is the final arbiter of truth, is the cornerstone of science.

The development of the scientific point of view had a great impact on all aspects of human thinking and social behavior, not to mention the growth of technology. If the authorities were wrong about something so basic as the location of the earth in the cosmos, then maybe they are wrong in some of their other claims. Although the idea had been around for a century or so, it now began to be more widely recognized that any claim, no matter how ancient and sacred the text in which it is written, is based on premises that are always open to question. In a short time after Galileo, the Age of the Enlightenment came to the West as poets, artists, and musicians explored the expanded horizons of open minds. While this was not the direct result of science—the flow of ideas most certainly was both ways, with science benefiting as well as contributing—there is no doubt that the astounding successes of the application of the methods of science must have left a deep impression on all thinkers of those times.

One of the most basic premises from time immemorial had been the concept of the divine right of kings. No matter how tyrannical or ineffectual a king might be, he was always strongly protected against overthrow by the belief that, for whatever mysterious purpose, God had placed him on the throne. Naturally each king did everything he could to promote that idea, actively discouraging the killing of enemy kings while holding no such compunction in regard to either his or his opponent's subjects. Then, in the seventeenth and eighteenth centuries, the concept of divine right came to be questioned. Kings lost their heads in England and France. Another ancient Greek idea, that people should rule themselves, was reborn and put into practice in France, America, and elsewhere.

How is it that the impact of great scientific ideas goes far beyond science itself, or even the technology produced from these ideas? Since it deals with the world around us and the basic questions of the origin and structure of the world, science ultimately affects all human endeavors.

It is somewhat ironic that Galileo was punished for teaching that the earth moves around the sun, which was not his idea but that of Copernicus and others before him. If he had not been tried for teaching Copernicanism, he probably would not get so much credit for its ultimate triumph. Galileo became the first astronomer of the modern age when he turned an artificial instrument, the telescope, on the sky. But in my mind, his truly unique contribution came as the first physicist of the modern age, since he initiated the development of the mechanics of moving bodies—the point at which physics and much of the rest of science begin.

Although Galileo's troubles were mainly theological, he was con-

fronted with one argument against the Copernican view that was legitimately scientific, and so had to be answered scientifically: it was said
that the earth cannot possibly be moving at the high rate of speed
implied by the Copernican theory (30 kilometers per second in modern
units). If it were, we would surely notice it. You can always tell when
you are moving, it was argued. Riding on a horse or a cart is certainly
a different experience from standing on the side of the road. Common
sense distinguishes between motion and rest. Common sense tells us
that the earth is at rest.

   Let us adopt this point of view and consider an experiment (fig.
2.4): A rock is dropped from the top of a building (the Leaning Tower
of Pisa?). Say it takes exactly one second to reach the ground. In that
time, if the earth is moving at 30 kilometers per second the tower will
have moved 30 kilometers away while the rock presumably falls straight
down. Thus the rock should hit the ground far from the base of the
tower. Of course it actually lands right at the base, which "proves" that
the earth is at rest.

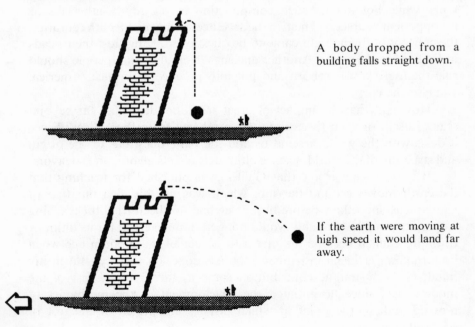

A body dropped from a
building falls straight down.

If the earth were moving at
high speed it would land far
away.

Fig. 2.4. "Proof" that the earth is at rest.

Galileo's answer to this apparent paradox was a profound one. First, the earth's motion is so evident from telescopic observations that it cannot be denied; this motion is an observational fact. Neither can one deny the fact that the rock lands near the base of the building. And if we do not notice the earth's motion in normal experience, then that too is a fact that must be reconciled. Thus, however we describe the motion of bodies, our explanation must be consistent with these facts.

Galileo recognized that, to describe motion, we must distinguish between *velocity*, which is a measure of how fast a body is moving, and *acceleration*, which is a measure of *changes* in a body's motion. When we ride a horse our body surely detects an aspect of motion, but that aspect is acceleration, not velocity.

The idea that we cannot sense our own velocity is much easier for people in the age of jet travel and space flight to comprehend than it was for those of seventeenth-century Europe, where a carriage ride of a few miles over the rutted roads between towns could be an exhausting experience. By contrast, hardly any sensation of motion is present as we cruise at 35,000 feet in the pressurized cabin of a large jet aircraft. The occasional jostle we feel when the plane hits some bumpy air, or the slight vibration from the engines, are indeed changes in our motion, that is, acceleration.

Similarly, when we stand on the earth, moving smoothly around the sun at 30 kilometers per second, we do not notice it. Just as our bodies show no effect of this high velocity, neither do the other bodies with which we share the earth. The rock dropped from the building behaves as it would if the earth were at rest. Galileo concluded that there is no distinction between being at rest and being in motion at constant velocity. In other words, all velocities are relative. We call this the *Principle of Galilean Relativity*.

The Principle of Galilean Relativity summarizes the basic observation that there is no intrinsic difference between the state of rest and the state of *uniform motion*—motion in a straight line at constant speed. If we are in a train moving along a straight track, we can imagine ourselves at rest while the countryside viewed from the window moves by. This is what is meant by *relative motion*. Most of us have had the disconcerting experience of sitting in a train in the station when an express train speeds by without stopping, creating a sudden sensation that we are moving.

Galileo's proposal that motion is relative was a fundamental break with the teachings of Aristotle (384–322 B.C.E.), whose ideas about motion had been accepted for centuries. Aristotle had assumed that a mover is necessary for motion, with God the *prime mover*. St. Thomas Aquinas (1224–1274) had used this as the first of his five "proofs" of the existence of God: "Whatever is moved is moved by another . . . it is necessary to arrive at a first mover, moved by no other; and this everyone understands to be God" (*Summa Theologica*, Q.2 Art. 3).

Even after Copernicus took the earth from the center of the universe, this concept of an absolute space framework was retained. Newton believed in it, though it puzzled him deeply. It was not until the twentieth century and Albert Einstein's reformulation of our concepts of space, time, and matter that the notion of absolute space was finally discarded. In hindsight, however, it seems that Galileo's discovery of the relativity of motion already contained the seeds of the destruction of the concept of absolute space.

The Principle of Galilean Relativity implies that no observation we can make, no experiment we can perform, measures an absolute velocity; any measurement of velocity is that of one body with respect to another. A car's speedometer measures velocity relative to the roadway. An aircraft's airspeed indicator measures its velocity relative to the air mass in which the aircraft moves, which itself may be moving at another velocity with respect to the earth. And, of course, the earth is moving at 30 kilometers per second around a sun that is moving through a galaxy, which is moving toward the Virgo supercluster of galaxies, and so on. But in all that motion, we still are at rest from our own point of view.

If we were to wake up inside a windowless room out in space, like the astronaut in the film *2001—A Space Odyssey*, we would have no idea how fast we were moving. No experiment we could perform inside the room would detect the motion of that chamber. Only by looking outside and watching stars go by could we measure any velocity, and only then with respect to those stars. Suppose the room really were at rest with respect to absolute space. How would we know it? We would not.

Absolute space is a concept invented by the human mind. We see that it is a flawed concept, one that fails the test of experiment. It follows then that it is meaningless even to speak of absolute space since such a concept can have no bearing on the outcome of anything that happens. What is the point of retaining a concept that cannot have any useful meaning, any unambiguous definition? The repeated verification of the Principle of Galilean Relativity in every observation leads us to conclude that the intuitive idea of an absolute space framework is simply another of the tentative ideas of primitive peoples, like that of the flat earth, which we must discard as we progress in knowledge of the universe.

The Principle of Galilean Relativity has withstood 400 years of experimentation. It faced a serious challenge at the beginning of this century, and the story of its rescue by Einstein is the theme of a later chapter. But first we need to follow the development of physics in the seventeenth century, from Galileo to Newton.

Isaac Newton (1642–1727) was born on Christmas Day in the year Galileo died. Perhaps the greatest mind in history, he showed unusual abilities even as a child. In one incident during his early teens, he won a contest to see who could leap the farthest, despite his poor athletic ability, by carefully timing his jump to take advantage of a gusting wind. He was a great experimentalist as well as theorist; blessed with remarkable eyesight, he could make laboratory observations that were impossible for others. Thus his achievements were not the product of thought alone, but of thought about the real world as it was presented to his senses. He explained color and gravity. He invented calculus and a telescope. But most important of all, he laid the foundation of science with the laws of motion and mechanics in the *Mechanical Principles of Natural Philosophy*, referred to as the *Principia*, published in 1687.

Drawing upon the observations of Galileo about the motion of bodies, Newton set forth three laws of motion. The first, the *Law of Inertia*, specified that a body left to itself will not change its state of motion: a body at rest remains at rest unless acted on by an outside force. The Law of Inertia also says that a body in uniform motion in a straight line will remain in that state of motion, unless it is acted on by an outside force. Common experience leads us to incorrectly conclude that a force is necessary for motion. An object pushed across a table will stop moving as soon as you remove your hand. Cars need fuel to keep moving. But a projectile tossed in the air keeps moving until it hits the ground, and the earth keeps moving around the sun with no one pushing it.

In a way, the Law of Inertia follows from the Principle of Galilean Relativity. Obviously a body at rest needs a force to get it moving. But the Principle of Galilean Relativity tells us that there is no difference between being at rest and being in uniform motion. So the same rule should apply in that case as well, since whether a body is moving or at rest depends on the observer's point of view. As with so many of the principles that describe the order of nature, this seems obvious once you think about it.

Newton's *Second Law of Motion* essentially defines the mass of a body as a measure of its inertia, that is, its reluctance to change its state of motion. It is easier to change the state of motion of an ant than an elephant, so the elephant has a greater mass. But although the elephant is also bigger than an ant, mass is not primarily a measure of size. A pillow may be bigger than a brick, but a brick will not be as easily stopped by a window. To stop something that is moving involves overcoming its inertia just as much as to start something moving that is initially at rest. Both cases involve acceleration, and acceleration requires a force. The Second Law is usually stated in terms of the famous equation $F = ma$, where F is the force that must be applied to a body of mass m to give it an acceleration a.

Newton's *Third Law of Motion* says that for every action there is an equal and opposite reaction. You kick a door open, and it kicks you back with the same force. This idea of recoil is more fundamentally associated with the notion of *momentum*. What we call momentum was defined by Newton as the *quantity of motion*. It is the product of the mass and velocity of a body, $p = mv$, where p is the momentum, m is the mass, and v is the velocity. On a football team, fullbacks have high mass so they cannot be easily stopped. Defensive linemen have high mass so they cannot be easily moved. Defensive backs need to accelerate quickly, and so have lower mass. But a low mass defensive back moving at a high velocity can still have sufficient momentum to stop a higher mass fullback running at a lower velocity.

Today we usually state the Third Law in terms of what is now recognized as one of the most fundamental principles in nature, *Conservation of Momentum*: the total momentum of a system is constant unless acted on by an external force. The First Law also follows from this basic principle, and the Second Law can be more generally stated, following Newton, as defining the force on a body as equal to the time rate of change of its momentum.

Newton drew a picture of the universe as a vast mechanism that operates under natural laws. The earth, moon, and planets behave in accordance with these laws and, in a break with previous belief, Newton proposed that bodies on earth obey the same laws as those in the heavens. This is most profoundly illustrated in his inference of the Law of Universal Gravitation. Measurements of the acceleration of falling bodies near the earth give a constant 9.8 meters per second increase in velocity each second. Newton realized that the moon falls around the earth just as a projectile falls to the ground. If you fire a shell horizontally from the top of a building it will follow a parabolic path to the ground. But if you could fire it fast enough, at least 7 kilometers per second, the curved earth would fall away as the shell traveled over the horizon and, except for the drag of the atmosphere, the shell would never hit the ground.

Using estimates of the distance to the moon known at that time and the fact that the moon takes about a month to go around the earth, Newton calculated that the acceleration of the moon divided by the acceleration of a falling body near the earth is in inverse square proportion with their respective distances to the center of the earth. Thus he concluded that the force responsible for the acceleration—gravity—obeys an *inverse square law*. Using his inverse square law, Newton was able to derive the laws of planetary motion previously discovered by Johannes Kepler (1571–1630).

The inverse square law was clearly an idea whose time had come. There is a story that Christopher Wren (1632–1723), the famous architect who was originally an astronomer, Robert Hooke (1635–1703), and Edmund Halley (1656–1742) each more or less independently also arrived at the inverse square law. They met together in a coffeehouse one day and discovered this, but could not figure out how to prove it. In August 1684, Halley went up to Cambridge and asked Newton what the curve would be for the paths of planets acted on by an inverse square law. Newton immediately answered, "An ellipse." Halley asked how he knew this, and Newton answered, "I have calculated it."

Hooke and Newton were mortal enemies, each believing the other had stolen his ideas. However, Halley had Newton's confidence and was able to convince him to publish his results in the *Principia*.

Halley is most famous for describing the orbit of the comet that now bears his name, predicting its reappearance 76 years later in 1758. Although Newton and Halley did not live to see it, the return of Halley's Comet exactly as predicted was a sensation that went far in establishing the credibility of the new science in the public mind. As I write this, Halley's Comet has just made another turn around the sun, the third reappearance since 1758, and the twenty-eighth appearance since it was recorded by the Chinese in 239 B.C.E.

The Newtonian mechanical framework as presented in the *Principia* and later reformulated by others is quite complete. Given the position, mass, and velocity of a body at a given instant, and the equation for the force acting on the body, we can use Newtonian mechanics to predict the motion of the body. That is, we can predict exactly where that body will be at some later time.

The ability to predict the future has always been a most sought-after prize. The motions of celestial bodies were used by ancient priests to predict the seasons, and this gave them great power. But these predictions were always imperfect. Now, with the mechanics of Isaac Newton, the world witnessed the growth of a new priestly caste, scientists, whose power to foresee the future far exceeded what anyone had had before. The ability to calculate precisely the motion of bodies has some obvious practical applications, such as in military ballistics. Less obvious is the way mechanics is used in the design of bridges, aircraft, engines, and electronic devices. In all modern technology, the starting point is Newton's mechanics.

The life of everyone on earth has been dramatically affected by the technology that is the outgrowth of Newtonian mechanics. Yet perhaps of even greater significance is the way mechanics changed the view of the universe. The mechanical universe is populated by particles moving about in accordance with the laws of physics, their behavior completely predictable. Everything in the universe is composed of these particles, and so everything in the universe behaves in a determined way. If human scientists lack the ability to calculate precisely when a hurricane will hit a coastline or an earthquake will shake a city, it is only because science has not yet learned how to make the calculations. The events still happen because they are destined to happen. It is written, if not in Scripture, then in the coordinates and velocities of all the particles and the forces acting on them.

This became the prevailing view, which lasted until this century. Of course people still held out hope that they, as human beings, were somehow different. It is still widely believed that life is more than particles moving along prescribed paths, that living matter has some special ingredient that makes it live, and that mind is something beyond matter. However, no evidence for this has ever been found. Plant, animal, and human tissue is composed of the same stuff as inorganic matter, although generally arranged in a more complex way. Chemicals affect the mind, producing "altered states of consciousness" or "separate realities." The "demons" that caused some people to become mad are now exorcised by drugs—clear evidence that thoughts and emotions are physical in nature. The line between living and nonliving has not yet been found; stopped hearts are restarted, breathing is performed mechanically, and the bodies of "brain dead" people are kept functioning by machines.

If human beings are made of the same particles as the rest of the universe, and those particles behave in predictable ways, then all of people's actions and behavior are determined ahead of time. There is no free will, and nothing in the Newtonian universe happens by chance. If certain things appear disorganized and unpredictable, it is because the details still remain to be discovered. A toss of dice is in practice unpredictable because we still are not smart enough to calculate all the complicated forces involved as the cubes twist and turn through the air. But in the Newtonian world view, we should someday be able to do it, as we should be able to predict earthquakes and human behavior. Nature, then, is completely orderly. Disorder only results as a false perception in the imperfect human mind.

We shall see later how this view was turned upside down by twentieth-century quantum mechanics. Human beings are still understood to be made of the same particles as the rest of the universe, interacting by the same forces. They are still machines, but quantum mechanical ones, subject to the vagaries of chance and often behaving in an unpredictable manner. Despite this, the Newtonian world machine still represents the most commonly held concept of the way in which science describes the world. The notion of a universe governed by perfectly deterministic natural law has been deeply implanted in our minds by the Judeo-Christian tradition.

Newton believed that the laws he was uncovering were set down at the creation by the same Lawgiver who presented Moses with the

Ten Commandments. He even doubted that God intervened to perform miracles: God is perfect and would have thought of everything ahead of time when He created the laws, and thus would never need to intervene to correct an error. However, it is the observation of nature, rather than the study of scripture, which is our path to the mind of God. Newton believed he was blessed as the person whom God had chosen to see His mind most clearly. Newton's successors, the scientists who had the learning and mathematical skills to penetrate the magical symbols of this new faith, then became the new priesthood, with powers of prediction and control over nature far exceeding any in history.

3

Reduced to Atoms

No single thing abides, but all things flow
Fragment to fragment clings; the things thus grow
Until we know and name them. By degrees
They melt, and are no more the things we know.

Lucretius, *De Rerum Natura* (paraphrase by Mallock)

The priesthood of science flourished only after Newton, and it gained its greatest powers over human affairs in the twentieth century. But it has roots more ancient than the many priesthoods it now discredits and supplants. If one were to pick a date for the birth of science, it might be May 28, 585 B.C.E. On that spring day about 2,500 years ago, in Ionia on the coast of present-day Turkey, Thales of Miletus is said to have used methods that we would still call scientific to successfully predict an eclipse of the sun. Can you imagine the wonder of the people of that time? It is said that the event stopped a battle.

The story is probably apocryphal, but surely an eclipse occurred that day. The exact day of every solar eclipse, and many other astronomical phenomena, can be calculated thousands of years in the past and thousands of years in the future. Just as scientific methods can be used to predict the future, they can also tell about the past.

The Babylonians and Egyptians had developed mathematics as a practical art—an important step but not quite science. Thales is also said to have visited Egypt and learned about geometry. While there, as the story goes, he determined the height of a pyramid by measuring its shadow at the time when his own shadow was equal to his height. This was a significant new step, because the concept of mathematically calculating something that cannot be directly measured is quite abstract.

This idea still distinguishes science from the practical arts. Science employs mathematics in an abstract, general way to describe phenomena in the natural world. A widely misunderstood point even in the present day is that science is far more than the sum of its practical applications. Scientists do not always press the point, since these applications form the primary basis for its financial support. Science experiments, theorizes, and generalizes, and then turns its results over to engineers, entrepreneurs, and politicians.

Another idea that became central to the development of science and has generated continuing controversy is also attributed to Thales. He is said to have been the first to propose that everything in the universe can be reduced to fundamental matter. He even went so far as to say that there was no basic difference between living and dead organisms, that they were all made of the same material. The seeds of the idea that life evolved from inorganic matter—still argued about even today— can be found in these thoughts. The science of nature initiated with these methods and ideas was called by the Greeks *physis*.

The notion that the fundamental material is particulate in nature is attributed to Democritus (late 5th–early 4th c. B.C.E.). He said that these fundamental particles were invisible and indivisible. They were called *atomos,* which means "uncuttable" in Greek, or *atomus* in Latin.

Unfortunately for scientific progress, Aristotle did not like the idea of atoms. We have already seen how the great Greek philosopher was wrong in his notions of absolute space and motion, probably delaying the development of science. Aristotle also argued against the atomic theory of matter. He could not see how the atoms could stick together to form objects. The rejection of the atomic theory by Aristotle was part of his and Plato's broader rejection of the scientific method, in particular its emphasis on observation and experiment.

Two major highways of philosophy emerged from ancient Greece. Science travels the original road straight out of Miletus, viewing the universe as a purely material entity, growing out of a primordial chaos, and operating automatically by way of the interaction of its basic elements. Over this road, science carries the fruits of the technology that grow in the rich soil alongside. A bypass around the road from Miletus began with Pythagoras around 550 B.C.E., but most of the paving was laid in the next century by Socrates, Plato, and Aristotle. These philosophers reintroduced the concept of a mystical or spiritual structure to the universe that lay beyond the senses. And this became the philosophical line adopted by Christian Europe in the following millennia. Even today, much of the western world travels this bypass; mystical belief in an undetectable reality still guides much intellectual and social behavior.

The mystical bypass curved back through the dark forests occupied by primitive superstitions. On the surface there may appear to be little connection between the mathematics and geometry of Pythagoras or the logic of Socrates, and the pantheon of gods behind every rock and tree that we associate with the pagan world. However, in believing that an invisible reality controls all events, Pythagoras and his successors moved their followers back into a dark cavern of thought from which rare light was glimpsed for another 2,000 years. The primary reason: an accident of history involving a Greek Jew named Saul.

Saul—Saint Paul—grew up in the Greek city of Tarsus. Speaking and reading Greek, he could not help but become aware of the dominant Greek philosophical notions of the day. So it is easy to understand how in his writings Greek mysticism and an acceptance of preordained fate became incorporated into Christianity, which was to become the dominant force in the western world. And when medieval theologians, particularly Augustine and Thomas Aquinas, rediscovered the Greek philosophers, it was natural that they should put them to use as ancient authorities to support and legitimize their own mystical theologies. In the meantime, Thales and the scientific method were largely forgotten.

The Miletus school survives today as science, thanks in large measure to the lasting influence of the materialist Epicurus (341–270 B.C.E.) and one of the great poems of the ancient world, *De Rerum Natura* (On the Nature of Things), written by the Roman Lucretius about 50 B.C.E. Epicurus denied the reality of anything that cannot be sensed or that produces any effect that cannot be sensed—the basic precept of science. Lucretius preserved the Epicurean tradition, including the idea of atoms, for future generations.

We have seen how Plato explained the observed wandering motion of planets as a disortion of their "true" circular paths, a largely useless explanation. The path you actually see is the one you want to be able to predict, and Plato did not tell how to do that. In science, the true path is the path that you see, because it is the only one you know anything about.

It is important not to leave the impression that it is necessary to see something visually to know it exists. As Lucretius noted, we cannot see the wind, yet it exists: "There are unseen particles of wind, since in their acts and ways they are discerned to rival rivers, whose substance can be seen" (*De Rerum Natura*, Book I verse 295, tr. Copley). Today we believe in atoms, and the elementary particles inside them, not because we can see them with our naked eyes but because of their effects, which can be seen with our eyes or by the magnifying lenses of modern instrumentation. And nothing that we observe with modern instruments shows evidence for anything in the universe beyond these elementary particles and the structures they combine to form.

As argued by Lucretius, there is evidence for only two "forms of being": *matter* and the *void*. If a third form of being were revealed, it would have to be *something*. If it could be touched, it would add to the ranks of matter. If it were intangible, unable to prevent the movement of a body through it, then it would be part of the void. And if there are accomplishments—art, music, literature—that set humans apart from the animals, these are nonetheless the results of the movements of atoms within the void.

It is not immediately obvious that matter is composed of discrete particles, separated by regions of empty space that are far greater than the size of the particles themselves. The solids, liquids, and gases of our normal experience look smooth and continuous. While the atomic theory of matter was revived by the mechanistic philosophers in the seventeenth century, it was not finally established to the satisfaction of skeptical scientists until 1905. Nevertheless, the evidence was always there for all to see, if only they had read Lucretius: "For watch whenever the bright rays of the sun pour shafts of light into a darkened house . . . you'll see many motes propelled by unseen forces, changing direction, turning, bounding back, whirling now here, now there, now everywhere. And this, you know, is how all atoms move" (*De Rerum Natura*, Book III, verses 115, 130, tr. Copley).

The random motion of dust motes in a sunbeam is evidence for the atomic nature of matter. This motion, when subtracted from that which results from air currents, can be attributed to the constant bombardment of the dust particles by the atoms in the air (fig. 3.1). This phenomenon is now called *Brownian motion*, after the botanist Robert Brown (1773–1858) who in 1827 observed the same effect for pollen grains in a liquid.

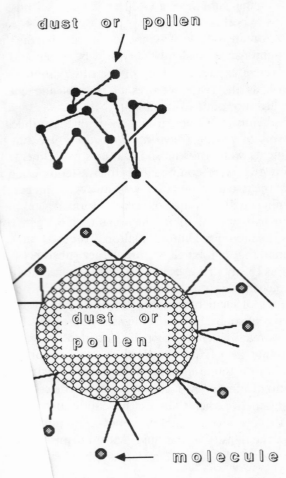

tion. The random motions of dust or pollen result from
particles by smaller molecules in the surrounding medium.

d by his observations. He very carefully isolated
nvironment, getting rid of all extraneous currents,
llen grains never stopped their random motions.
ad discovered the fundamental particles of life,
vital forces, which at that time were believed
ter. However he was able to observe similar

effects in nonorganic subtances and, like the good scientist that he was, abandoned this hypothesis.

While Brownian motion was not totally ignored in the nineteenth century, its significance was only very slowly recognized. The atomic theory nevertheless explained many of the observations being made in the laboratory, as the sister sciences of physics and chemistry began to stride forward, not quite in step.

Modern chemistry was born in 1808 with the publication of John Dalton's *A New System of Chemical Philosophy*. By that time eighteen chemical elements were known, and Dalton's experiments had revealed that when any two were combined to form a compound, the ratio of the weights of the two amounts used was always an integer. For example, when combining hydrogen and oxygen to form water, the ratio of the weight of the hydrogen used to the weight of oxygen equaled seven, at least as measured by Dalton. Dalton proposed that the element were basic units that combined to form other substances, but he woul not go so far as to say that these atoms were separate material particle

While nineteenth-century chemists studied how the various elemen and compounds of elements reacted together, physicists were studyi other properties of solids, liquids, and gases with results that the chemi did not always readily believe. In 1811, Amedeo Avogadro (1776–18 hypothesized that at a fixed temperature and pressure, equal volur of gases contain the same number of particles. This implied that a could be viewed simply as a collection of particles bouncing aro inside its container, rather than as the continuous smooth fluid appears to the eye, with equal volumes containing different m depending on the density of the fluid. Most chemists, including Da refused to accept this; fifty years later it was still being debated at chen conferences.

It certainly is no great fault for a scientist to be skeptical; i it is an important part of the job. The chemists were right in dema to be convinced, not by fancy equations but by quantitative experir results. These were slow in coming, but the beautiful simplicity particulate picture of matter in explaining many phenomena cou be denied.

It often happens in physics that an elegantly simple idea gain ents because of its theoretical beauty, even while experimental cc tion is slow in coming. Other sciences, such as chemistry, ter more conservative, which may help keep physicists honest but also sometimes deters these others from making grand revolutionary leaps in the understanding of nature. To see how physicists came to the atomic view, we need to go back to the previous century.

After the American Revolution, a Royalist American, Benjamin Thompson (1753–1814), thought it better to move to Europe. For helping the King of Bavaria fight a war, Thompson was rewarded with the title Count Rumford, after his hometown in New England. While boring cannon for the king, he noted that the barrels became hotter when the boring tool revolved faster. This suggested to him that heat is connected to motion, and he later presented a paper to this effect before the Royal Society in London. Previously heat had been thought to be a physical substance, *caloric*, which moved from body to body. This proved to be wrong. You can easily demonstrate the connection between heat and motion by rubbing your hands together and feeling the warmth generated. The quantitative relationship between heat and various forms of mechanical energy was later verified in a series of experiments by James Joule (1818–89) in England. We now conventionally measure energy in units called *joules*.

In the meantime, Daniel Bernoulli (1700–82) had proposed that the pressure a gas exerts on the walls of its container is the result of collisions of particles with the walls. These observations led to the development of the *kinetic theory* of gases (fig. 3.2), primarily by Rudolf Clausius (1822–88), James Clerk Maxwell (1831–79), and Ludwig Boltzmann (1844–1906). Boltzmann would later commit suicide, despondent over the nonacceptance of his ideas. In the kinetic theory, the temperature T of a body is simply a measure of the average kinetic energy of the particles in the body, when expressed on an absolute temperature scale for which $T = 0$ corresponds to zero kinetic energy.

The *internal energy* of a gas is the total energy of its particles, including any rotational or vibrational energy of the particles themselves. When one body is in contact with another, heat flow from the hotter body to the colder one occurs as the transfer of internal energy. The *First Law of Thermodynamics*, which requires that any heat given off by a body be compensated either by work done on the body or by a lowering of internal energy, then follows as a consequence of what is now recognized as one of the most fundamental principles of physics, *Conservation of Energy*. In the absence of any external work being

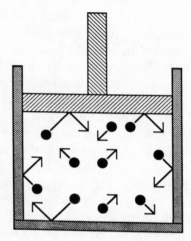

Fig. 3.2. In the kinetic theory of gases, pressure in a container of gas results from the collisions of molecules with the walls.

done on or by a system of bodies, the total energy of the bodies in the system, whether kinetic or potential, remains constant.

The atomic theory, applied to both chemistry and the physics of gases, liquids, and solids, gradually gained adherents. But one major flaw legitimately kept it from achieving total acceptance: there was no direct experimental measurement of the sizes of the individual atoms or the number of them in a particular body. Dalton's atomic theory said that there are two hydrogen atoms for every oxygen atom in water (H_2O), but did not specify how many atoms there are in a body. Are there 100, 1,000, 100,000? How big are they? What do they weigh? No one had firmly established the scale of atoms.

Some remarkably good estimates of the size of atoms were made in the nineteenth century based on laboratory observations of a wide variety of phenomena in optics, thermodynamics, electricity, and magnetism. By about 1890, these observations had converged on a consensus that a typical atomic diameter was about 10^{-8} centimeter. Since one cubic centimeter of solid matter weighs about a gram, it was inferred that the atom must have a mass of the order of 10^{-24} gram. Looking at this another way, one gram of matter must contain about 10^{24} atoms.

This particular quantity—the number of basic particles in one gram of a substance—is called *Avogadro's number*. It is now known more precisely to be 6.022×10^{23} particles per gram. Incidentally, for this purpose it turns out that the basic particles are not what we now call atoms, but rather the protons and neutrons that make up the nuclei of these atoms.

Despite the wide consensus of the value of Avogadro's number, the reality of atoms was still not conclusively established by these estimates. Then, on April 30, 1905, a patent clerk in Bern, Switzerland, named Albert Einstein completed his Ph.D. dissertation, "On a New Determination of Molecular Dimensions," and submitted it to the University of Zurich. In this work, Einstein (1879–1955) showed how measurements of the Brownian motion can be directly applied to determine the value of Avogadro's number. The idea was simple and just like that given by Lucretius in the passage quoted above. Others, notably Jean Perrin in France, had realized the significance of Brownian motion and were making careful laboratory measurements; Einstein's important contribution was to derive the exact equations needed to relate the measured data to Avogadro's number.

The random movement in Brownian motion results from the constant bombardment of the visible pollen or dust particles by much smaller invisible atoms or molecules. Just like the wind, which we do not see while it produces observable effects, these atoms and molecules produce real effects that we can observe and measure. We can easily understand how measurements of the random motions can give an estimate of Avogadro's number. If atoms were smaller than they are, Avogadro's number would be bigger than it is, because it would take more atoms to constitute a gram of matter. In that case, we would observe a smoother motion of the Brownian particles resulting from a greater number of smaller impulses from these tiny atoms. On the other hand, if the atoms were bigger than they are, and Avogadro's number smaller, there would be fewer hits with greater impulse, producing more ragged, disjointed motion. Einstein related the disjointedness of the motion directly to Avogadro's number and other measurable quantities.

With Einstein's work and the data from experiments by Jean Perrin and others, Avogadro's number was determined and found to be completely consistent with the earlier, less direct, estimates. By 1910 or so, opponents of the atomic theory of matter had either died or been converted. Today, the concept that matter is mostly empty space and composed of great numbers of fundamental objects is as firmly accepted as any in science. Thales, Democritus, Lucretius, and Epicurus are finally vindicated.

4

You Can't Trust Common Sense

Newton, forgive me; you found the only way which in your age was just about possible for a man with the highest powers of thought and creativity. The concepts which you created are guiding our thinking even today, although we now know that they will have to be replaced by others farther removed from the sphere of immediate experience, if we aim at a profounder understanding of relationships.

Albert Einstein

The atomistic view of the universe as a vast collection of infinitesimal bodies in motion is an ancient idea. Yet, when we look thoughtfully at the world around us, we become aware of phenomena that seem to be something other than material. Once convinced that wispy clouds and bubbling rivers are really groupings of vast quantities of atoms, we still are strongly urged by our intuition to regard a beam of light as yet another distinct physical entity.

A special place has always been reserved for light as a fundamental component of nature, different from material bodies. In the biblical Book of Genesis, God's first words are, "Let there be light." Once there is light, then the material universe is created. In science, the study of the unique nature of light ultimately has led, in the twentieth century, to two great revolutions of physics and human thought: Einstein's Theories of Relativity and Heisenberg's development of quantum mechanics.

Newton is usually credited with the hypothesis that light is in reality a stream of particles. He actually was a bit more circumspect: "Tis true, that from my theory I argue the corporeity of light; but I do so without any absolute positiveness . . . I knew, that the *properties,* which I declar'd of *light,* were in some measure capable of being explicated not only by that, but by many other mechanical hypotheses" (Cohen, 1958). In other words, light may or may not be composed of material bodies, but it is undoubtedly mechanical.

Newton had observed the way a prism splits a ray of white light into separate rays with the colors of the rainbow. Clearly light was composite—composed of rays of different colors. Eighteenth-century atomists promoted the corpuscular theory of light, with Newton as their authority. This was atomism carried to its ultimate extreme, with light not fundamentally different from the rest of the substance of the universe.

An alternate theory of light had been proposed by Christian Huygens (1629–95) in Holland. Huygens suggested that light was a vibratory phenomenon, like ocean waves or sound waves. Huygens's wave theory was not immediately accepted because of the greater authority of Newton, but by the nineteenth century it had been adopted as the correct explanation of light for the best of reasons—it agreed with observations. The eighteenth-century atomists who so fervently promoted the corpuscular theory of light were replaced by scientists who promoted the wave theory with equal fervor.

Newton had objected to the wave theory of light because, he said, light does not bend around corners like sound. Even his exceptional eyes missed a few things; it is a simple matter to demonstrate that light does indeed bend around corners. Just take an index card and punch the tiniest hole you can in it with a pin. Then hold the card up to a light (for safety, do not use the sun or a laser beam). You will see the light diffuse around the hole, giving a spot which appears much bigger than the hole itself. The light from the source does not travel straight through the hole, as might be expected if it were a stream of particles, but rather spreads out in passing through. Surprisingly, the smaller you make the hole, the more the beam spreads out.

A more precise experiment can be done with a laser source and a circular aperture, using as a screen a section of blank wall rather than the fragile retina of your eye. You will see a bright central spot surrounded by alternate bright and dark rings. And if you try different colored lasers, you will observe that the diameters of the central spot and rings will depend on the color, the diameter being greater for red than for blue. This phenomenon, called *diffraction*, is neatly explained by Huygens's wave theory of light as the constructive and destructive interference of the wavelets at the aperture. The phenomenon can be simply understood by analogy with water waves, which will show the same effect (fig. 4.1). By the nineteenth century many careful laboratory

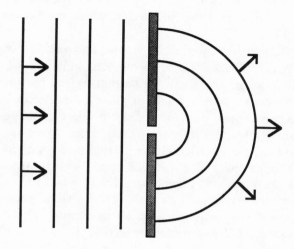

Fig. 4.1. In the phenomenon of diffraction, light or sound waves bend around corners.

experiments, very difficult to conduct without the benefit of lasers, had demonstrated that light experiences the diffraction and interference effects normally associated with waves. Ultimately, these led to the recognition that light was closely related to two other phenomena—electricity and magnetism.

In the nineteenth century, the phenomena of electricity and magnetism were put to marvelous application through the efforts of Michael Faraday (1791–1867) and many others. Then in 1867 occurred one of those great moments of insight that happen perhaps once a century: James Clerk Maxwell wrote down four equations that described all that was known about electricity and magnetism at that time. Bits and pieces of these equations had previously been discovered by Faraday, Gauss, and Ampère, but Maxwell showed how they all fit together into a complete picture. In particular, he illustrated that the electric and magnetic fields are really manifestations of the same fundamental force.

Maxwell noted that the equations of Gauss, Faraday, and Ampère were not quite symmetric, so he added a term to make them more symmetric. Faraday had earlier demonstrated that a changing magnetic field produces an electric field, a fact used in today's electric generators and motors. Maxwell's new term predicted that a changing electric field will produce a magnetic field, an effect that had not yet been observed in the laboratory.

Once this term was inserted in Maxwell's equations, they admitted another solution. Previously, the incomplete equations of electromagnetism required an electric charge or current to feed energy into electric or magnetic fields. Maxwell's equations showed that electric and magnetic fields can exist in a vacuum where there are no electric charges or currents to feed them. Rather, they feed on each other. Once started, energy oscillates between the electric and magnetic fields. Further, these fields propagate through space at a speed computed exactly from the known constants of electricity and magnetism. These constants are determined from experiments having nothing directly to do with light, yet the speed at which Maxwell's electromagnetic oscillations move is exactly the speed of light (c)—c = 300,000 kilometers per second. Thus Maxwell discovered that light is an electromagnetic wave moving in a vacuum as well as through transparent media such as air or glass.

From the type of diffraction experiments described earlier it had already been determined that the *wavelength* of visible light—that is, the distance from crest to crest in the wave—ranged from 4×10^{-7} meter for the blue end of the spectrum to about 6×10^{-7} meter at the red end. However, Maxwell's equations did not forbid other values for the wavelength. In the novel *The Once and Future King* by T. H. White there is a fortress entered by tunnels in the rock. Over the entrance to each tunnel is written a line that has become a cardinal principle in physics: "Everything not forbidden is compulsory." When the existence of some phenomenon not generally observed in nature is not specifically ruled out by known principles, then one should look for it or find out why it is not there. Maxwell's equations predicted the existence of electromagnetic waves of other wavelengths than those of visible light. Such waves had never before been observed. Then in 1886 Heinrich Hertz (1857–94) was able to generate radio waves with wavelengths of the order of one centimeter, confirming Maxwell's prediction and commencing the world of almost instantaneous communication that we have today.

Important aspects of science and the scientific method are illustrated by this story. We must continually remind ourselves that the principles of science are based on observation. Nevertheless, new discoveries of a fundamental nature are often the result of insights motivated not so much by experiment or observation but by belief in an underlying beauty and simplicity. It is truly a marvel to write Maxwell's equations on the blackboard and manipulate them around until electromagnetic waves almost magically appear. That so few people are capable of appreciating the beauty and the grandeur of this result is a pity. The electric and magnetic fields and the equations of electromagnetism are very abstract, requiring a knowledge of vector calculus which only those studying physical sciences or engineering ever obtain. Thus even highly educated graduates with advanced degrees in other fields lack the tools to appreciate and comprehend the phenomena which probably have a greater effect on them than any other, since electromagnetism pervades every aspect of our existence.

Maxwell's equations imply the existence of electromagnetic waves. Waves are usually understood as vibrations of some physical substance, such as air or water. So it is natural to imagine that electromagnetic waves are also vibrations of some kind of material medium. However, since light travels to us through space from distant stars the medium of vibration must pervade all the universe. It must, in some sense, be the stuff of space itself.

By the end of the nineteenth century most scientists believed that the universe was contained in an invisible, frictionless medium called the *ether* invisible because it could not be seen, frictionless because the heavenly bodies moved endlessly through the ether without being retarded. Lucretius would have been puzzled by this, since in his view the "form of being" that did not prevent the movement of bodies through it could not be matter but only the void.

If the ether pervaded all of space, it had to form the absolute space framework in which all the bodies of the universe moved. When a body was at rest with respect to the ether, it was absolutely at rest. When it moved with respect to the ether, it was absolutely in motion. The reference frame of the ether could be singled out as a very special one. If all this were so, then the Principle of Galilean Relativity, which says that there is neither absolute rest nor absolute motion, could not be true. This is not just an abstract argument; it can be tested experimentally. In 1887 the American physicists, A. A. Michelson (1852–1931) and E. W. Morley (1838–1923), set out to do just that and, as usually happens, they obtained results that they did not anticipate.

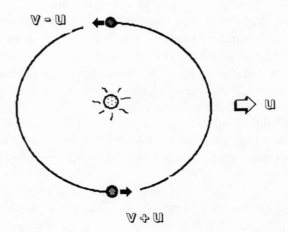

Fig. 4.2. The motion of the earth through the "ether." Assume the solar system is moving through the ether with a speed of u. The earth moves in orbit around the sun with a speed v. Depending on the time of year, the speed of the earth through the "ether" changes from $u + v$ to $u - v$.

Their experiment has been described many times, so let us just give the idea here. Imagine yourself in a boat moving along on the ocean. Waves slap against the hull as they move toward shore. If the boat is moving away from shore the waves will hit the boat at a high rate; if you turn the boat and move toward shore the waves will hit the hull at a slower rate. With your eyes closed you can sense the direction of the boat by the sound of the waves slapping against the hull and, in fact, measure your speed relative to the water.

Michelson and Morley tried to do the same thing, with the earth as their boat and the ether as their ocean. They built a device that measured the difference in the "slap" rates of light beams, projected at right angles and then returned by mirrors to an eyepiece. The precision of their instrument was more than sufficient to distinguish the change in light speed of 60 kilometers per second that would be expected to occur as the earth circles in its orbit around the sun (fig. 4.2). With many trials over months, and different orientations of the apparatus, they never found any measurable difference in the slap rate of the two beams of light. Unlike water waves, whose speed depends on how fast a boat moves through the water and the direction of motion, light waves move at a speed that does not seem to care about the motion of the light source or the observer. Michelson and Morley failed to find evidence for the ether.

If Michelson and Morley had done their experiment in 1600, it might have been regarded as experimental evidence that Copernicus and Galileo were wrong. The results are what one would expect if the earth were at rest in an absolute space defined by the ether. Remember, however, that the best argument against the idea that the earth moves is the fact that we do not sense its motion in everyday experience. This fact led Galileo to introduce the idea that velocity is relative. In the Principle of Galilean Relativity, there is no observation you can make that allows you to sense your velocity, without using some outside reference point. When you drive along a smooth, straight highway at constant speed, only the passing countryside gives you a sense of motion— a very dangerous situation, which has led the designers of modern highways to build in gentle curves even across flat plains, to remind drivers they are moving at high speeds relative to the objects with which they might collide.

In failing to observe the earth's motion with their self-contained apparatus, Michelson and Morley actually verified the Principle of Galilean Relativity. If they had found that the speed of light depended on the motion of the earth, they would have discovered a violation of relativity; the result would have implied a detectable difference between rest and motion. So, far from showing that Galileo was wrong, Michelson and Morley gave significant additional testimony that Galileo was right. And in the process they discovered experimentally what we now recognize as a consequence of the validity of the Principle of Galilean Relativity: the speed of light in a vacuum is always $c = 300,000$ kilometers per second, independent of the motion of the source or observer. Incidentally, Maxwell's equations are consistent with a fixed speed of light in a vacuum, a theoretical result that probably had more to do with what followed than Michelson and Morley's experimental result.

The minds of men and women are often enclosed in a veil which prevents them from recognizing the obvious. To lift that veil may require a person of exceptional genius. Today we regard it as obvious that the earth is round, yet it was once thought to be flat. Today it is obvious that the motion of everyday bodies is relative, yet the genius of Galileo was required to recognize this. To the modern physicist the relativity of space, time, and mass is obvious, but the extraordinary genius of Albert Einstein was needed to show how the relativity of these familiar concepts is an inescapable consequence of the relativity of motion.

Albert Einstein was born in the city of Ulm, Germany, on March 14, 1879. Like Newton and so many of the great minds in history, Einstein's origins were humble; his father made featherbeds. And, like Newton, Einstein showed early signs of exceptional ability. The stories about him being a poor student are simply false. He was generally first in his class, although his innate distaste for arbitrary authority did not endear him to all his teachers as he matured. Thus he had a difficult time finding work after graduating from the University of Zurich in 1900. Eventually he landed a position in the patent office in Bern, the capital of Switzerland.

Between reviewing patent applications Einstein found time to continue his physics studies. Then, in perhaps the most remarkable single year of creativity in human history, Einstein was transformed from a nobody to the man of the century. By March 17, 1905, before being officially awarded his Ph.D. degree, Einstein had written a paper that would win him the 1921 Nobel prize. In it he proposed that light is composed of particles called photons, a concept that, as we will see, helped trigger the quantum revolution. We have discussed Einstein's dissertation on Brownian motion, which helped to confirm the atomic theory of matter. The paper based on this work was received by the editors of the journal *Annalen der Physik* on May 11, 1905. On June 30 they received his first paper on relativity. His paper proposing the famous equation, $E = mc^2$, arrived on September 27. And finally, on December 19, a second paper on Brownian motion was submitted. In a few short months Einstein had achieved more than any scientist in history except Newton, and it took Newton a lifetime. We now turn our attention to Einstein's *Special Theory of Relativity*, the first of the two revolutions of twentieth-century physics.

We have seen how western culture, through the all-pervasive influence of the Judeo-Christian tradition, has deeply implanted in our minds Aristotelian concepts that are wrong because they disagree with unprejudiced observations of the world around us. Even the most highly trained scientist is not immune to accepting concepts that have been passed down as dogma, although those concepts may be contradicted by the facts. This is doubly true when a concept seems, at first and even second glance, to be common sense.

In America, especially, we have great faith in common sense. We are exhorted to apply it to every activity; so many problems would be solved if only everyone applied "good old-fashioned common sense." Galileo and Newton had shown that written authority carries no special corner on truth. Their lead was followed in the French and American revolutions, which discarded the authority of kings, but this was replaced with the authorities of Thomas Paine's "common sense" and Thomas Jefferson's "self-evident" truth.

Common sense can be wrong, and self-evident truth should be no less open to question and doubt than written authority. It was once self-evident truth that the earth was at rest at the center of the universe. It was once common sense that bleeding was a way to cure disease, by letting out the poisons that are the cause. It was once common sense that animals, or people, had to be sacrificed to soothe the anger of the gods.

Until Einstein, it was self-evident truth that the concepts of space, time, mass, and energy were immutable properties of the universe and its constituent bodies. Einstein found that they were not, that they depend on one's point of view. What you measure for the length or the mass of a body depends on how fast you or the body is moving. If that body has a clock attached, the clock will be observed to run slower than a clock held in the hand of the observer. If you measure the body's mass, it will be greater than when the body is at rest. And the same for energy, since energy and mass are equivalent.

Einstein recognized that space, time, and mass-energy are *relative*; that is, they depend on the relative motion of the observer and the thing being observed. He determined this by going back to what we have already noted as the most basic phenomenon in the universe: the motion of bodies. Einstein then straightforwardly examined the logical consequences of the relativity of motion first elucidated by Galileo.

Rather than follow Einstein's mathematical presentation, let us use a simple but elegant thought experiment usually attributed to Nobel Prize-winning physicist Richard Feynman. Suppose one makes a clock out of two parallel mirrors, between which a light pulse bounces back and forth (fig. 4.3). A counter on top of one mirror keeps track of the ticks by moving the arrow on a dial one notch each time the pulse hits the mirror. Suppose that the mirrors are 15 centimeters—about six inches—apart. Then each tick will be 10^{-9} second or one nanosecond, the time it takes light to travel the 30 centimeters from one mirror to the next and back.

Now set the clock moving at a high speed in a direction parallel to the mirror surfaces, and take a time-exposure photograph of the moving clock. Of course all this is hard to do in practice, but technical difficulties will not deter us from concluding what the result would be if we had the resources to do the actual experiment. The photograph will show the light pulse tracing a diagonal path from the bottom mirror to the top one, as the top mirror moves ahead in the time it takes the light to reach it. You might ask how we know that the light will keep hitting the mirrors once it's moving. Remember relativity: the clock is moving in your reference frame but not in the frame of an observer moving along with it. To this observer the clock is at rest and the light pulse simply bounces up and down, hitting the mirrors just as it does for the clock in your reference frame. The light *not* hitting the mirrors would violate the Principle of Galilean Relativity.

We have seen that the speed of light cannot depend on the motion of the source or observer. The diagonal path we observe for the light pulse is greater than the distance between the mirrors, so the light appears to take a longer time to get there than when the clock is at rest. The

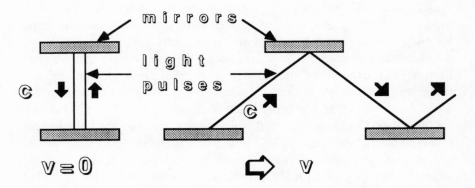

Fig. 4.3. Time dilation. The light pulse clock on the left is observed to be at rest. The clock on the right is observed to be moving with a speed, *v*. Since the light pulse always moves with the same speed, *c*, it takes longer to travel the diagonal path between mirrors. Thus the moving clock appears to run slower.

hand on the counter that displays the ticks will move around the dial at a slower rate than it does when the clock is at rest. The moving clock apparently slows down! Time intervals measured on the moving clock will thus be shorter than on the clock at rest. From a simple application of the Pythagorean theorem,* the slowing-down factor for a speed v can be shown to be the square root of $(1 - v^2/c^2)$. For example, let $v = 0.6c$, that is, the clock moves at 60 percent of the speed of light. Then the slowing down factor is square root $(1 - 0.6^2) = 0.8$. The observer might measure ten minutes between two events on his own clock, while the moving clock, running slower, reads only eight minutes. Time, at least as measured by a light pulse clock, is relative. We call this effect *time dilation.*

A common objection at this point is that we have demonstrated only that light pulse clocks run slower when they are moving. Throwing an alarm clock against the wall will also make it run slower, but that does not let you catch a few more winks without being late for work. The boss's watch keeps absolute time, as far as he is concerned. This kind of relative time is not what we are talking about here. Today we have time standards that are accurate to tiny fractions of a second. Atomic clocks can be made that provide pulses able to repeat at intervals as small as 10^{-12} second, one *picosecond*, or less.

Suppose we send an atomic clock, along with a light pulse clock of comparable precision, with an astronaut in a spaceship. We watch the spaceship fly by the earth at high speed and observe the light pulse clock slow down, as before. Suppose the atomic clock does not similarly run slower. The astronaut then observes a strange effect. To her, the atomic clock runs slower than the light pulse clock. The difference will cease when the spaceship is again brought to rest. But if this happens, then we have again violated the Principle of Galilean Relativity: we would have found an experiment capable of distinguishing between absolute rest and absolute motion. A body would be absolutely at rest when all clocks read the same as light pulse clocks. When they read differently, then the body would be absolutely in motion. If we require the Principle of Galilean Relativity to be valid, then all moving clocks, not just light pulse clocks, must be observed to run slower.

---

* Pythagorean theorem: for a right triangle, the square of the hypotenuse equals the sum of the squares of the other two sides.

When we say that all moving clocks appear to run slower, this includes any clock, no matter how complicated its construction. The human body is such a clock, with its biological processes ticking along at some rate. If biological clocks were somehow different from physical clocks—if they were made of different stuff—we would then have a way to measure our motion, and Galilean Relativity would be wrong. If we make the working assumption that Galilean Relativity is correct, then we are forced to conclude that the astronaut in the spaceship will have her body clocks running slower. She will appear, to those of us left on earth, to age more slowly.

When first confronted with time dilation, students frequently complain, "I can't understand it." What they really mean is, "I don't believe it." This is a legitimate complaint, because the relativity of time is not common sense; it is not an everyday experience because the effect is very small at the speeds involved in normal experience. For example, take astronauts in orbit above the earth. Their speed relative to the ground is perhaps seven kilometers per second, a high speed by earthly standards yet only about two thousandths of one percent of the speed of light. Clocks on board the spacecraft lose only 3×10^{-10} second each second, a measurable but hardly noticeable loss. On earth we move around at even lower speeds with respect to one another and simply do not notice these small differences in time. So time dilation is not common sense, but it does happen, as has been repeatedly verified experimentally in the years since the predictions by Einstein.

We have already noted that modern atomic clocks can measure time to accuracies better than a picosecond, a trillionth of a second. In 1971 J. C. Hafele of Washington University in St. Louis, Missouri, and Richard E. Keating of the U.S. Naval Observatory flew atomic clocks on regularly scheduled around-the-world flights, both east to west and west to east. When the clocks were compared with similar ones left on the ground, the eastward clocks had lost an average of 59 microseconds, while the westward clocks had gained 277 microseconds. These results were quantitatively in agreement with the predictions of both special and general relativity.

In the realm of particle physics, where accelerators produce particles with speeds as high as 0.9999999997 of the speed of light, time dilation is an everyday laboratory occurrence that has been confirmed innumerable times by accurate measurements. The high energy cosmic rays that hit the earth's atmosphere also demonstrate time dilation. Unstable particles called *muons*, produced in the atmosphere by cosmic ray interactions, reach the earth in far greater numbers than they would if their lifetimes were not extended by their high speeds. Every one of the hundreds of muons that pass through our bodies every hour is mute testimony for the relativity of time. Life on earth might not have developed as it did without this background radiation providing an occasional mutation to nudge evolution along. Our very existence is perhaps a consequence of time dilation.

Needless to say, the idea that a moving clock runs slower than one at rest was initially met with a certain incredulity. People intuitively found the idea of relative time difficult to believe, and many scientists looked for holes in the argument. The most famous "hole" that Einstein had to plug was called the *twin paradox*.

Suppose there are two identical twins, one of whom goes off in a spaceship. According to Einstein, she will age less than the sister left behind and, when she returns, will now be younger. But speed is relative. To the twin in the spaceship, her sister on earth is moving at a high speed and aging more slowly. So when they again meet, it should be the earthbound twin who is younger. In other words, time dilation leads to the paradoxical situation where one person is younger from one point of view, and the other person is younger from another point of view. How can this paradox be resolved?

There really is no paradox when one looks carefully at the experiment. The points of view of the twin sisters are not equivalent. The astronaut twin must accelerate to reach the high speed, accelerate again in turning around out in space, and then decelerate to return to the earth's reference frame. The twin left on earth remains always at rest with respect to the earth.

Recall that acceleration is quite different from velocity. When we say that motion is relative, we specifically mean velocity, the rate of change of position. Acceleration, the rate of change of velocity, can be detected without recourse to external observation. The astronaut in the spaceship can measure her acceleration with a simple accelerometer: a mass on a spring. Or, even more simply, she will be pushed back in her chair as the ship increases its speed.

This does not prove, of course, that the astronaut returns the younger of the two. But it does show that there is no paradox because the frames of reference of the two sisters are not equivalent. Let us study the situation further, because the result will be important in our later cosmological discussion.

The problem of acceleration in relativity is a tough one. Einstein had to use the mathematics of non-Euclidean geometry to solve this problem in his *General Theory of Relativity*, first published in 1916. We can avoid this in discussing the twin paradox by considering a different experiment, which involves constant velocities but still must yield the same results as when the twin is accelerated.

Suppose that instead of one spaceship we have two, and that at midnight on January 1, 2000, one of these spaceships passes close to the earth, moving at 60 percent of the speed of light. At this instant the astronaut on board synchronizes her clocks with the ones on earth. She travels outward from the earth for 8 years, as measured on her clock, until at midnight on January 1, 2008, she is passed closely by another ship heading back toward earth at exactly the same speed. The astronaut on board this second ship sets his clock to the time and date signaled to him by the first astronaut. He continues back to earth at constant speed, arriving back at midnight on January 1, 2016, as measured on his clock.

Now let us consider the passage of time back on earth. The clocks in each spaceship will be observed to run 80 percent as fast as an earth clock. Or, equivalently, the earth clock will run faster by a factor of 1.25. Thus, while 16 years elapse on the moving clock, 20 years will have passed as measured by clocks back on earth.

This experiment gives the same results as the accelerated twin paradox experiment. If we assume that the astronaut twin's spaceship can accelerate to $0.6c$ very quickly, then she can ride alongside the first ship going out, turn around quickly, and ride alongside the second ship coming back. Her clocks will read the same as those in the two ships. We can draw the conclusion, which comes out of the more precise mathematical arguments of general relativity, that *an accelerated clock runs slower*.

Before discussing these ramifications, let us think further about the twin experiment. Again we have an example which defies common sense. How can two identical twins ever be anything but the same age? This assault on common sense becomes even harder to swallow if we raise the velocity of the spaceship. Suppose it is $0.9998c$ and the ship leaves earth at midnight on January 1 of the year 0 C.E., traveling 20 years out to a star before turning around and taking 20 more years to return, for a total trip of 40 years. At that speed the ship's clock appears from earth to run slower by a factor of 50, so 50 × 40 = 2,000 years elapse on earth and the ship returns at midnight on January 1, 2000, as measured on clocks on earth. A 25-year-old astronaut could have left the earth at the time of Christ and returned in the year 2000, *at an age of 65 years*! As difficult as this is to accept, it is the clear implication of relativity.

A common response at this point is to say: "All you have shown is that moving clocks read differently. But clocks are mechanical things built by people. Time is a property of the universe and your clocks simply do not give the correct true time."

This is our old nemesis Aristotle talking. The assumption made is that time is something inherent to the universe with an existence all its own, independent of space, matter, or humanity. In fact, time is an invention of the human brain to help describe and classify the data of the senses. Quite simply, as we have already seen, *time is what is measured on a clock*. Without the universe, without matter, without a clock to measure time, time has no meaning. What we mean by the intervals of time between events in the universe are just those things that are measured on clocks: the number of ticks of a watch, the number of bounces off a mirror, the number of turns around the sun, the number of heartbeats of a human being. Any and all of these constitute clocks.

Any type of clock could be used as a standard of time, if everyone agreed to use it. We might imagine a despotic ruler decreeing that his heartbeats should be the time standard. Think of the trouble that would cause! Time would depend on his condition each day. Suppose a laborer were contracted to work for 24,000 heartbeats a day. He might work from dawn to dusk on a day when the king rested; another day, when the king jogged and played a game of basketball, the laborer might finish by noon. Needless to say, the description of even the simplest physical phenomenon, such as a falling body, would be extremely complicated by such a standard, since the king's activities would have to be factored into every equation. In our scientific world we use more objective standards, not because they are any more correct than subjective ones, but because they afford us simpler descriptions of natural phenomena. They are more practical, but are still arbitrary.

The daily rising of the sun, the monthly phases of the moon, and the regular reappearance of the seasons have provided us with our common standards of time: the day, month, and year. But these were eventually found to be inadequate. The ancient Babylonians corrected calendars in which the year was a fixed number of days. They found astronomical observations a better way to predict the seasons, as did the Egyptians and the builders of Stonehenge. Today's astronomers use sidereal time as a standard, but still need complex tables to make corrections for the precession of the equinoxes in order to predict precisely the direction to aim their telescopes to find a particular body. Astronomical time is better than biological time, but still not good enough.

In 1967, an international agreement was made that the second would be defined as 9,192,631,770 periods of the wave of light emitted from a particular atomic transition in ^{133}Cs, an isotope of cesium. This was about 1,000 times more accurate than astronomical standards, or one part in 10^{12}. Undoubtedly this will not be the last standard, as we continue to make strides in our ability to measure the minute changes in physical phenomena, which we describe within the framework of space and time.

Just as time is what is measured on a clock, *space is what is measured with a meter stick*. Lay down a meter stick between two points in space: the number you read off the scale is the distance between them. The primary international unit of length is the *meter*. Once the meter was defined as the length of a particular platinum-iridium bar kept under strict temperature control in a laboratory in Paris. A few years ago, however, it was finally accepted that length is really not something independent of time. Now the meter is defined in terms of the distance traveled by light in a fraction of the standard second. That is, once the second is standardized, you have no need for a new standard for length. The speed of light in a vacuum is the same in all reference frames, so the distance between two points can be measured by a light beam and a clock that records the time the beam takes to go between the points.

It follows that if time is relative, space must also be relative. Recall the astronaut who traveled at $0.9998c$ to a star. The trip there and back took 2,000 years as measured on an earth clock, and since the astronaut's speed was almost equal to the speed of light (c), the star must have been 1,000 light-years from earth. But wait—the astronaut aged 40 years during the trip, 20 years on the way out and another 20 years on the way back. As far as she is concerned, the star is 20 light-years from earth. To her the distance of 1,000 light-years from the earth to the star has contracted to a mere 20.

Imagine that distance to be laid out with meter sticks end to end; these will each have been contracted so that the same total number still stretch out to the star. Just as the astronaut's clock runs 50 times slower than an earth clock, to an observer on earth two points in space one meter apart on earth would be contracted to 2 centimeters in the eyes of the astronaut. This effect is called *Fitzgerald-Lorentz contraction*, after the two physicists who independently conceived of the idea prior to Einstein as a possible explanation for the Michelson-Morley experiment. Space, as well as time, is relative. But then that should be true, if space is really defined in terms of time.

Once we have established the relativity of time and space, we find that mass too must be relative. Newton had introduced the concept of mass as a measure of a body's resistance to changes in its motion. Einstein noted that this mass will also be observed to increase as a body is accelerated to high velocities. We can crudely think of it this way: mass is a measure of the inertia or sluggishness of a body. Suppose you watch a body go by at a high speed. All the particles in the body will appear to move around more slowly because of time dilation. Remember each is a little clock. So the body will appear to behave more sluggishly, just as if its mass has increased.

This result led Einstein to another profound conclusion: he realized that the increase in mass of a body is, except for units, exactly what we call *kinetic energy*. Historically, energy had been regarded as a separate concept from mass because it was believed that the mass of a body was fixed while the energy of a body changed. Indeed, mass came out a constant in all laboratory measurements. When chemists measured the mass of two substances, mixed them together and watched them react, and then measured the total mass of the resulting substances, including any gases produced, they consistently found that the total mass was unchanged. They called this *Conservation of Mass*. But mass is not conserved in fact; these measurements were simply not accurate enough, or not done with particles moving fast enough.

Much is made, and properly so, of Einstein's famous equation $E = mc^2$. But c is just a constant that makes the units come out right: *kilograms* for mass and *joules* for energy, in the metric system of physical units. This equation simply, but nonetheless profoundly, states that mass and energy are the same thing.

Once again we find a common-sense notion cast into the heap by Einstein. It had been thought that the universe was composed of two kinds of stuff: *matter* and *energy*. Material bodies are matter; light and other forms of radiation are energy. Einstein told us that there is really only one kind of stuff. In his first 1905 paper, he had restored the Newtonian concept of light as composed of material bodies, which he called *photons*. These photons carry energy from one point in space to another. Thus they also have an inertial mass. One usually hears that photons are massless, but this refers to *rest mass*, the mass measured when a body is at rest. Since a photon has never been observed at rest, but rather always moves at the speed c, this rest mass has never been directly measured. Indirect limits place it at less than 10^{-33} of the mass of the electron, the lightest known particle that has a rest mass.

So it is not surprising that the particles of light and the particles of matter were thought for so long to be separate types of stuff. With Einstein's Special Theory of Relativity these two types of stuff, matter and energy, were unified. As we will repeatedly find as we continue our story, the march of physics is toward an ever-increasing simplicity in our description of the universe.

5

The Hubble Bubble

I was sitting in a chair in the patent office at Bern when all of a sudden a thought occurred to me: 'If a person falls freely he will not feel his own weight.' I was startled. This simple thought made a deep impression on me. It impelled me toward a theory of gravitation.

Albert Einstein

When Einstein published his General Theory of Relativity in 1916, he held to the traditional and still prevailing view that the universe is a stable, unchanging place. It was known from geological studies that the earth was not always as it now appears. It formed, evolved, and still evolves, as the giant crustal plates continue their movement and volcanos create new land. The heavens, however, seem unchanging. The apparent motion of the stars was thought to result from the earth's motion through the firmament—a word that suggests a fixed eternal framework within which things move. Despite the wide acceptance of the Principle of Galilean Relativity, which says that all motion is relative, Aristotelian ideas of absolute space still persisted in our thinking about the universe. So when Einstein's equations predicted that the stars and galaxies would be found to be moving, either away or toward one another, he added a *cosmological term* to the equations to stop this motion. This term implied an additional, possibly repulsive, aspect of gravity, which is not present in the Newtonian theory, to stabilize the universe.

If we fire a body straight up in the air, it will eventually fall back to earth, unless its initial velocity exceeds the *escape velocity*, 11 kilometers per second. Galaxies can be thought of as bodies moving in the gravitational fields of all the other bodies in the universe. Even with Newton's concept of gravity we expect that the galaxies will be moving toward or away from other galaxies and that a static unchanging universe with each galaxy at rest is highly unlikely.

Einstein later called his introduction of the cosmological term the biggest blunder he ever made, for it was shortly discovered by Edwin Hubble (1889–1953) that the galaxies are indeed receding from one another at a rate of speed that, as well as observations can tell, is proportional to their distances. The proportionality factor, called the *Hubble factor*, is consistent with no cosmological term; the galaxies are simply moving away from one another as if the universe were expanding like a vast balloon or bubble. Later we will see that Einstein's "blunder" is possibly no error after all. The cosmological term appears to be negligible now, but there may have been a time when, for a brief fraction of a second, the universe rapidly inflated under the action of a repulsive cosmological force.

In his General Theory of Relativity, Einstein developed a new law of gravity that replaced Newton's and had great implications for our understanding of the nature of the universe. First he proposed that there really is no *force* of gravity; what we observe in nature as gravity is basically the deflection of the path of a body in the presence of another body. In Newton's physics, this requires a force, since the natural path of a body is in a straight line in the absence of an external force. In Einstein's new physics, the observed deflection away from a straight line—as in the case of the elliptical orbit of a planet around the sun— results from the body following its natural path, which is no longer a straight line. Einstein arrived at this conclusion when he attempted to bring acceleration within the framework of relativity.

We have referred frequently to Galileo's recognition that it is impossible to distinguish one's velocity without outside information. We noted, however, that this is not the case for acceleration. Inside the closed cabin of an aircraft you are pushed back in your seat as the plane accelerates down the runway, and forward against your seatbelt when the plane brakes to a stop during landing. Another common experience of acceleration occurs when we are traveling on a curved path. Turning to the right, we imagine we feel an acceleration to the left. Turning left we experience an acceleration to the right.

In the seventeenth century, Christian Huygens had introduced the term *centrifugal force* for the outward push a body experiences when it moves in a circle. In those days it was believed that circular motion, like that of the celestial bodies, was natural, so an agent, or force, was needed to change that natural motion. Newton countered that the situation was the opposite: the natural motion of a body was in a straight line, and a force was required to cause the body to deviate from that straight line. Newton said that a *centripetal force*, directed toward the center of a circle, was needed to keep a body moving in that circle. For example, if one swings a ball on the end of a string in a circle, the tension in the string provides the centripetal force that keeps the ball accelerating toward the center of the circle. Similarly, it was believed that the planets were kept moving around the sun by a centripetal force that, unlike the string, could not be seen. This, according to Newton, was the force of gravity. Many were troubled by this invisible "action at a distance," but Newton's concept of gravity worked so well in explaining phenomena on earth and in the heavens that it was quickly and fully accepted.

In Newtonian mechanics, the inward centripetal force that keeps a body in a circle is the real force and Huygens's outward centrifugal force is fictitious. Centrifugal force is introduced so that the laws of motion can be applied in a rotating reference frame. When someone drives a car around a corner and feels himself pushed against the door, we say this push is the result of centrifugal force. Because we want to believe that a force is needed to produce an acceleration, we invent one! Once again, our subjective view of things leads us astray; we prefer to view things with ourselves at the primary reference point. An observer with a more cosmic perspective, viewing the car from a fixed point above the roadway, would see what is really happening: The body of the driver is moving naturally in a straight line, but the car turns, and the door pushing against him provides a centripetal acceleration to change his direction of motion.

Now Einstein noted the following. Suppose that a closed capsule is in space and accelerating at 9.8 meters per second per second. A woman awakes in the capsule with no memory of where she actually is. She finds herself standing on the floor. She drops her purse and it falls to the floor. Accustomed to spending her time on the surface of a planet where the acceleration due to gravity is 9.8 meters per second per second, she naturally assumes she is in a room on earth.

Suppose another person awakes inside an elevator near the top floor of an extremely tall building, just after its cable has snapped. He experiences himself floating around the elevator. He might think, "I

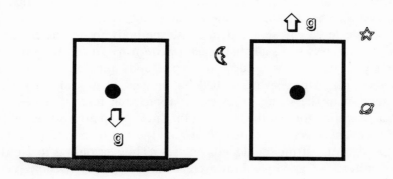

Fig. 5.1. The Equivalence Principle. On the left, a body is observed to fall with an acceleration *g* in a room on the earth. This observation is indistinguishable from one made in the room on the right, which is moving out in space with an acceleration *g*. Thus gravity is indistinguishable from acceleration, and the effects of gravity can be duplicated by the acceleration of reference frames without recourse to the gravitational force.

am in a capsule in space," before he is decelerated to a quick stop at the bottom of the shaft.

So if you are in a closed capsule and do not know whether you are on earth or in space, you have no way of telling. This apparent lack of distinction between gravity and acceleration is called the *Principle of Equivalence* (fig. 5.1).

Einstein recognized that Newton's invisible gravitational force could also be viewed as fictitious, like centrifugal force or the similar Coriolis force that we use to explain effects of the earth's rotation. He extended the Principle of Galilean Relativity to accelerated systems by noting that it is not only impossible to distinguish between motion at constant velocity and being at rest, but it is also impossible to distinguish between changes in the velocity—that is, acceleration—and gravitation. His original 1905 Special Theory of Relativity dealt with constant velocities. The 1916 General Theory of Relativity dealt with acceleration, but in the process became a Theory of Gravitation. And since gravity is the primary force at cosmic distances, it finally became a Theory of Cosmology.

In General Relativity the force of gravity is fictitious. In place of Newton's invisible force pulling a body away from its natural straight path,

Einstein proposed that the natural path of a body is distorted by the presence of matter nearby. A planet curves around the sun because that is its natural path, just as Huygens assumed. When there are no other objects around, a body moves naturally in a straight line, as Newton said.

Einstein was able to describe these natural motions by resorting to what up to that time had been regarded as a mere mathematical curiosity: *non-Euclidean geometry*. The axioms of Euclid and the theorems derived from them are among the wonders of the world. They are taught to high school students as an introduction to deductive logic. The philosopher Immanuel Kant used Euclidean geometry as an example of how knowledge can be obtained by pure thought, that is, without recourse to observation. He argued that these principles applied to the real world, yet could be derived by reason alone. He neglected to note that the points and lines of geometry are abstractions from the real world, and the principles of geometry were undoubtedly suggested to the ancient geometers by figures drawn in the dirt with a stick. Geometry, no less than any science, is based on observations of the real world.

Euclid's fifth axiom says that two parallel lines never meet; equivalently, the sum of the interior angles of a triangle is equal to two right angles. For centuries geometers thought it should be possible to derive this principle from the other four axioms, but all their attempts failed. In the early nineteenth century, Carl Friedrich Gauss (1777–1855) actually tried to test the fifth axiom experimentally by measuring the angles between beams of light flashed between mountain tops. He obtained agreement with Euclid, but recognized that this did not necessarily mean that the universe is Euclidean. His experiment merely showed that space is Euclidean for distances like those between mountains, within his experimental errors. Gauss saw the possibility of other geometries in which the fifth axiom is relaxed. Afterward the mathematical structures of these geometries were developed by Lobachevski, Bolyai, and Riemann. So when Einstein came along with the need to have parallel lines meet without a force to move them together, he found the mathematical structure already in place.

The surface of a sphere such as the earth is the example of a non-Euclidean geometry usually given in the textbooks (fig. 5.2). Two "parallel" longitude lines each make angles of 90 degrees at the equator, yet meet at the poles. The sum of the angles of the triangle formed by the two longitudes and the equator is greater than the sum of two right angles.

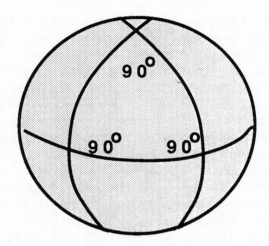

Fig. 5.2. In non-Euclidean geometry, "parallel" lines can meet, and the sum of the angles of a triangle can be greater than 180 degrees on the surface of a sphere.

Einstein wrote down a set of equations to describe the motion of a body in the presence of other bodies. In these equations the effect of gravity is present in the term that describes the way the geometry deviates from Euclidean: the greater the mass of a nearby body, the more the geometry becomes non-Euclidean in the sense that a particle follows a path of greater curvature. Most important, Einstein predicted phenomena that were not permitted in Newton's gravitational theory, as it was then understood.

The first of these was the precession of the perihelion of the planet Mercury—the axis of the planet's orbit rotates 1.5 degrees per century. While this is a tiny effect, the classical form of Newtonian gravity does not predict this. Einstein's calculations using General Relativity agreed with previous observations. This was a small triumph, because theories that explain previously known facts are always suspect. One wonders how much the inventor of the theory cooked it up to make the answer come out right, not necessarily out of dishonesty, but out of a strong human desire to succeed. The prediction of a phenomenon not previously observed and not expected from conventional ideas has far greater credibility. And if the prediction is quantitative—that is, if it provides a

numerical value to be checked agianst objective measurement—then we can judge the value of the theory on a basis of more than mere speculation and personal opinion.

The second prediction of General Relativity was that light passing near a massive body will be bent as it follows its natural path, curved by the body's presence. This effect had not been previously observed, and early in 1919 Arthur Eddington led an expedition to Brazil to attempt to verify Einstein's prediction. On May 29 of that year, a total eclipse of the sun would occur and stars in the background near the rim, made visible by the moon covering the sun, should appear displaced as their light was bent in passing the sun. An effect was seen, although a bit larger than the calculated 0.87 seconds of arc. It often happens that experimental errors and other complications prevent clear answers to questions at the edge of knowledge. Over the years, however, the observations of many more eclipses and other tests with radio waves and laser beams have verified that the path of a light ray is not always a straight line, and the deflections agree with the calculations of General Relativity.

Let us think more about this. What is a straight line? Suppose I want to cut a long straight tunnel through a mountain. How might I determine that the tunnel is straight? I could shine a light beam down one end and see if it is visible at the other. In another example, suppose I want to lay out a series of rocks in a straight line along the ground. How might I do this? As I place each one on the ground, I look down along them from the end and see if they all "line up." These examples clearly assume that light travels in a straight line to the eye. What if the light really travels in a curve? Then the rocks will line up only if they lie along that curve. Similarly my tunnel will be curved if I dig it using a curving light beam as a guide. Furthermore, I will not know it is curved. To me it will look straight!

This becomes a matter of semantics, as it always does when competing concepts fail to provide an empirical test to choose between them. This is the reason science has found it important always to insist upon operational definitions of concepts, and to conclude that when two concepts lead to the same observations, then the concepts are equivalent.

What we normally define as a straight line is actually the path followed by a beam of light. What Einstein tells us is that the path will change depending on the presence or absence of matter in the vicinity; whether that path is straight or curved is irrelevant. We have seen that time is what you measure on a clock, and that the distance between two points in space is the time that you measure on a clock for a light ray to go between the points. There is no mention of whether that ray travels a straight or curved path. In General Relativity the path followed by a light ray is called a *geodesic*, in analogy with the great circles on a sphere that represent the shortest distance between points on the surface.

One of the commonest misconceptions about General Relativity is that it shows that space is curved. Space is not curved, any more than time or mass are curved; neither is it straight. This is just a convenient way of speaking, which will be used later in this book. The precise statement, however, is the following: Bodies move on paths that can be described by a non-Euclidean geometry in the convention of General Relativity. But they can just as well, though perhaps not always as elegantly, be described by a Euclidean geometry. The only difference is whether gravity is regarded as a fictitious force or a real force. The General Theory of Relativity is a description of how bodies move in the presence of gravity. The theory made predictions of phenomena not present in the classical Newtonian picture using Euclidean space. But today we realize that these same phenomena also follow from purely Euclidean formulations, once Special Relativity is properly included.

For example, consider the bending of light by the sun. We will learn later that, despite the success of the wave theory, light is a stream of particles after all. We call the particles of light *photons*. These photons carry energy E, and therefore, from $E = mc^2$, they have inertial mass m. In fact, *energy is inertial mass* since c is just a constant, serving only to change from units of mass to units of energy when we feel compelled to work in archaic units, developed when people thought that mass and energy were different. In a totally Newtonian way, then, a photon will be attracted toward the sun and caused to deviate from its straight path by the action of the force of gravity on its inertial mass.

What we keep finding as science develops is that the principles we uncover are really not *laws of nature* but quite arbitrary inventions of the human mind. They develop by a series of almost accidental discoveries that work well and so gain attention and acceptance. Other ideas might work too, but once a particular line of thought is started, tradition and the weight of authority keep it going in a particular way until we reach an impasse. Then some exceptional person is able to break new ground and start things moving in another direction. We are learning that there is nothing wrong with keeping several independent lines of inquiry moving in parallel. The most popular theories are often those that are the most elegant and simplest to use, or sometimes just the ones that are in current fashion. The successes of one particular approach do not necessarily establish the invalidity of other approaches; only disagreement with observation can do that.

Finally, we reach the third prediction of General Relativity: *Light in a gravitational field will be shifted in color toward the red.* For example, the light from a star viewed near the edge of the sun during an eclipse should not only appear shifted from its normal position in the absence of the sun, but also appear redder than its normal color. Early attempts at verification were disappointing because of the difficulty of understanding all the possible conventional sources of a red shift of light.

As has been said, we cannot claim to have observed a new effect, even when the data are in agreement with the theoretical prediction, until all ordinary explanations are convincingly ruled out. Not until the 1960s were most scientists convinced that the gravitational red shift had indeed been seen, in astronomical observations or in the lab.

The most convincing evidence for the gravitational red shift was found in 1960 in laboratories at Harwell in England and at Harvard University in the United States. There experimenters actually measured the frequency shift of *gamma rays*—very short wavelength electromagnetic waves—as they fell down a tower. The precision needed for these experiments was tiny—one part in 10^{12}—but this was achieved.

We can understand the gravitational red shift if we return to our earlier discussion of the twin paradox. Recall that we were able to show that an accelerated clock will appear to run slower. In two steps we can demonstrate how the gravitational red shift follows. First, gravity and acceleration are equivalent, so a clock in a gravitational field will appear to run slower. Second, the vibrations of an electromagnetic wave constitute a clock, and so will be shifted down in frequency—in other words toward the red end of the spectrum. To explain further, we have already noted that our standard clock is based on the frequency of the electromagnetic vibrations of the cesium atom. So the frequency of the electromagnetic vibrations that we call light will be reduced in a gravitational field, like any clock. The red end of the spectrum of light is the lower frequency end, so the color of the light is shifted toward the red.

The gravitational slowing of clocks was also checked in the Hafele-Keating experiment discussed earlier. Cesium clocks were sent on board regularly scheduled airline flights, while other cesium clocks remained on the ground. The time differences measured were combinations of the time dilation of Special Relativity, resulting from the velocity of the aircraft and the earth's rotation, and the time dilation of General Relativity, resulting from the minute differences in the gravitational field on earth and in the air. All this was calculated over the complete flight path of the aircraft, and the measurements and calculations agreed within the expected experimental errors.

The development of General Relativity came at a time when astronomers were discovering that the universe is far larger than had previously been thought. Since the time of Newton, it had been suspected that the stars were other suns, much farther from us than our own sun. However, the first good estimates of distances to the stars were not obtained until the early nineteenth century, using the parallax caused by the earth's orbital motion. This technique worked only for the nearest stars. In 1912, Henrietta Leavitt of Harvard discovered a relationship between the pulsation period of Cepheid variable stars and their luminosity. Once this technique was calibrated with stars whose distances were known from parallax, it could be used as a way to measure the distance to Cepheids that were too far away for parallax.

At first Leavitt's papers were ignored, but in 1920 Harlow Shapley, working at Mount Wilson Observatory in Pasadena, California, used the technique to establish that the center of our galaxy of stars, the Milky Way, is some tens of thousands of light-years away. This implied that the galaxy was far larger than astronomers at the time believed. Shapley's result was met with skepticism, especially since he did some very subjective and unjustified smoothing of the data, but he was right nonetheless.

Astronomers for many years had observed other objects that did not appear as points, the way stars do, but rather as fuzzy blurs. These were called *nebulae*, and scientists had long speculated that some of these might be galaxies of stars like the Milky Way. Using the Mount Wilson telescope, Hubble was able to resolve individual stars, including Cepheids, in several nebulae, thus verifying that some were indeed galaxies and also obtaining a measure of their distances.

It happens that it is easier to measure velocity in astronomy than to measure distance because the light from stars contains very precise *spectral lines* in addition to the continuum of colors radiated by any hot object. The existence of spectral lines cannot be understood by the classical electromagnetic theory of Maxwell and was one of the observations which triggered the development of quantum mechanics. Basically spectral lines are a series of very pure colors that are characteristic of individual atoms and molecules. *Spectroscopy*—the observation and analysis of spectral lines—is a prime tool for determining the presence of the most minute quantities of particular materials in a sample of matter. It is also the method by which astronomers determine the composition of the atmospheres of stars and planets. Because of the *Doppler effect*, the shift in the frequency of observed spectral lines can be used to determine the velocity at which a body is approaching or receding from us.

Hubble had an assistant at Mount Wilson named Milton Humason who started out as a mule packer on the mountain, graduated to bus boy and then janitor, and eventually became a skilled and tireless observer. Humason spent many long, cold nights in the observer's cage measuring the spectra of galaxies, which Hubble then analyzed. Our closest neighbor galaxy, Andromeda, had earlier been found by others to have a blue shift, implying that it is moving toward the Milky Way at about 200 kilometers per second. Hubble and Humason, however, found that the majority of galaxies had red shifts, indicating that they are receding from us. Hubble used the Cepheid yardstick and discovered that the recessional velocities of the galaxies implied by the red shifts of the spectral lines of the stars in these galaxies are proportional to their distances to earth.

This relationship is called *Hubble's Law*: It is expressed as $v = Hd$, where v is the velocity, d is the distance to the galaxy, and H is the *Hubble factor* already mentioned (fig. 5.3). This gave Hubble a way to estimate the distance to galaxies too far away for individual Cepheids inside the galaxy to be seen, assuming the proportionality held true at those distances. One measures the velocity v from the red shift and divides by H. Hubble obtained a value for H that was a factor of ten or so higher than current estimates. Making quantitative estimates is often difficult in astronomy. In this case, it turned out that at least two kinds of Cepheid variables exist, confusing the interpretation. Even today, the Hubble factor has a rather large uncertainty, estimates ranging

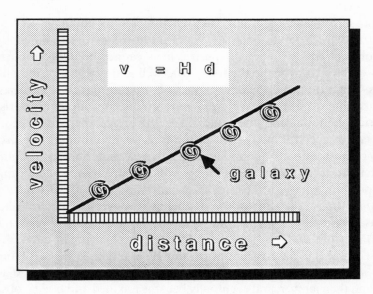

Fig. 5.3. Hubble's law shown in graph form. The distance of a galaxy, d, is proportional to its recessional velocity, v.

from 15 to 30 kilometers per second velocity for each million light-years of distance to a galaxy.*

If you had taken a consensus of astronomers on the size of the visible universe in 1916, the year of the publication of Einstein's paper on General Relativity, you probably would have obtained a figure of about 1,000 light-years. In the early 1920s, after Shapley's work on the Milky Way, this estimate would have increased to several tens of thousands of light-years. In the 1930s Hubble, with his verification that nebulae are galaxies and his discovery of the velocity-distance relation, estimated 100 million light-years; but he was low by a factor of ten. Galaxies have now been observed with red shifts corresponding to velocities that are a significant fraction of the speed of light. The distance implied by a velocity of 30,000 kilometers per second, one tenth the speed of light, is of the order of 1 billion light-years. Today we know that it is a few billion light-years from our solar system to the galaxies farthest from us.

---

* The conventional unit of distance in astronomy is the *parsec*, which is equal to 3.26 light-years. The Hubble factor thus is usually given as 50–100 kilometers per second per megaparsec, where a megaparsec is 1 million parsecs.

We have seen how, 400 years ago, Copernicus triggered an upheaval in human thinking by removing the earth from the center of the universe. The astronomy of the following centuries showed that our sun was just another star, but not until early in this century was it realized that we are located in an obscure corner of a galaxy of 100 billion suns, and that the visible universe contains hundreds of billions of such galaxies. The importance of these facts, which show the insignificance of human beings and the earth, is slowly creeping into the public consciousness. Most people find it difficult to grasp the reality of the numbers involved, a distance of a billion light-years or a quantity like 10^{22}, the number of stars in the known universe.

The pictures of earth taken from the moon by the Apollo astronauts have been perhaps the greatest recent contribution to public awareness of the cosmos. These showed us our world from outside, giving us the cosmic perspective of ourselves. What we saw was a blue planet covered with white clouds looking not at all like the maps studied in school, which usually show the United States in green and the U.S.S.R. in red. The continents can barely be discerned below the clouds, and no sign of the boundaries between nations or any human works can be seen. But even in these pictures, the earth still looks very big and the stars very small. In reality the earth could easily be swallowed up by our sun, and the sun is a comparatively small star. There are stars whose diameter would extend beyond the earth's orbit. There are other stars with ten times the mass of our sun, and when these reach the end of their lives, they explode in a flash that can be seen galaxies away.

So we have seen our perspective on the cosmos change drastically in this century; the universe is much bigger and more varied than ever dreamed. But perhaps even more significant has been the realization that the cosmos is a changing, evolving place. Stars are born out of clouds of matter and pass through a series of evolutionary steps before burning out all their nuclear fuel. They may end their lives as *white dwarfs*, *neutron stars*, or *black holes*, depending on their masses. Our sun is a loner, far from other stars. In many cases, however, stars are found close together in gravitationally bound systems in which matter is exchanged and each component passes through several evolutionary stages and back again.

Galaxies too are found to evolve. When we look at those galaxies with the greatest red shifts, we are looking at the way the universe was hundreds of millions of years ago, since that is the length of time it has taken the light of those galaxies to reach us. We find that these galaxies are different from closer ones. The most distant galaxies we can see, the *quasars*, radiate such enormous quantities of energy that they look like nearby stars that just happen to have an enormous red shift. The name *quasar* comes from the term *quasi-stellar object*. Even today a small group of distinguished astronomers still argue that these objects are not as far away as their red shifts imply. However, the majority have become convinced that quasars are galaxies that were formed long ago and so are at great distances. It is believed that the nuclei of these galaxies contain black holes with masses millions of times that of the sun. The huge radiation of quasars then results from disturbances at their cores, as the stars and other matter in the inner regions are swallowed up by these nuclear black holes.

Black holes are probably the most fascinating objects in today's astronomy. They too were a prediction of General Relativity, implied by Einstein's equations though not immediately recognized by him. As for much else in General Relativity, black holes can be crudely understood with Newtonian ideas; indeed the eighteenth-century French astronomer and mathematician Pierre Laplace (1749–1827) seems to have recognized the possibility. The *escape velocity* needed by a body to leave the gravitational pull of a planet or star is proportional to the square root of the ratio of the mass of the body to its radius. As has been noted, for the earth this is about 11 kilometers per second. Now suppose that all the mass of the earth were compacted into a sphere with a radius of slightly less than one centimeter. The escape velocity from earth would then exceed the velocity of light. Since no body can travel faster than light, it would be impossible for anything to escape, including light. The earth would have become a black hole.

The radius at which a body becomes a black hole is called the *Schwarzchild radius*. In principle, black holes can come in any size, but strong repulsive forces between the particles in bodies keep most from becoming black holes. Current theories of stellar evolution predict that stars about 2.5 times more massive than the sun become black holes at the end of their cycle, when all their nuclear fuel is exhausted. In that case the mutual gravitational attraction of all the burned-out nuclei in the star is so great that it overcomes the repulsive forces and collapses to a point called a *singularity*. Since no light is emitted or reflected, a black hole is the blackest imaginable object. But that does not mean that it cannot be detected.

The presence of a black hole is disruptive, as the black hole pulls in everything within reach. Atoms swallowed into the black hole are stripped of electrons and radiate X-rays and other particles that can be observed. A black hole can also be detected by the way its mass distorts the orbits of nearby bodies. Observations of these types of effects have led to a good indirect case that the binary star system Cygnus X-1 contains a black hole.

Evidence is also gradually accumulating for a super-massive black hole at the center of our own galaxy. Highly intense radio waves are seen from a region smaller than the solar system. Perhaps all galaxies have central black holes and started their lives as quasars, with their radiation diminished as the supply of matter has dwindled. We only see quasars that are far away and, because of the great time needed for light to reach us, are being viewed as they were long ago.

So the sky is not a "firmament," but rather a constantly changing scene that appears fixed only on the puny time scale of human beings. The human species did not even exist when the light rays we now see from most galaxies started their long journeys to earth. Hubble found that galaxies are receding from one another at a speed proportional to their distance. This result has a simple but exceedingly profound interpretation: the universe is expanding like the fragments of an explosion. The *Hubble law*, $v = Hr$, is exactly what one would expect if at some time in the past the universe had exploded. What we now see are galaxies moving away from one another, each with a velocity imparted in the original explosion. Those which happened to have higher velocities have moved farther away in the time since the original event, called the *Big Bang*.

If we assume that there was a range of velocities from zero to the maximum c, then the radius of this expanding shell of gas we call the universe must be c/H, or from 10 to 20 billion light-years. The Big Bang then occurred from 10 to 20 billion years ago. A long time, but not eternity.

The idea of the Big Bang has not been quickly accepted by science. Many alternatives have been proposed to explain the red shift: perhaps it is a gravitational red shift; or perhaps the light gets "tired," that is, loses energy as it makes its way over vast distances. A small but dedicated group of astronomers have still not given up the idea of a *steady state universe*. They claim that this is more consistent with the relativistic notion that there is no preferred position in space or time. The Big Bang seems to imply a special time, *t = 0*, when the universe began. However, the steady state theory requires the continuous creation of matter out of nothing to fill in the spaces voided by expansion. This violates one of the prime principles of physics, Conservation of Energy. While we should not rule out the possibility, we can show that Conservation of Energy follows from the very principles the steady state theory takes as its hypotheses. The theory is logically inconsistent.

But regardless of theory, there is strong evidence for the Big Bang. In 1964 two radio astronomers working for the Bell Laboratories in New Jersey, Arno Penzias and Robert W. Wilson, were trying to measure the radio waves produced outside the plane of our galaxy. They were studying that particular component of the static, or noise, which interferes with electromagnetic communication. What they heard turned out to be the sound of the Big Bang, still reverberating throughout the universe. Penzias and Wilson later shared a Nobel Prize for their discovery.

In 1932 the radio "window" to the universe had been opened by another Bell Labs scientist, Carl Jansky. As Penzias and Wilson were later, Jansky was being employed by Bell Labs to study the sources of radio static. After localizing several sources, such as lightning and thunder, he found that another source lay roughly in the direction of the center of the Milky Way. The implication was that radio waves are reaching us from a distance of 10,000 light-years. Since then radio telescopes have become powerful tools for astronomical observations. Astronomical bodies are not limited in their emanations to the tiny spectral band of visible light that humans happen to be able to detect with unaided eyes. The neutron stars mentioned earlier were first observed as radio pulsars. Radio pictures of the universe obtained today show magnificent vistas of galaxies shooting out jets of matter millions of light-years into space. And radio is providing us with evidence that there are organic-like molecules in the spaces between stars.

Penzias and Wilson observed a background of microwave noise that seems to be coming from all directions with uniform intensity. This intensity is low, about one ten-millionth of the power of a 100-watt light bulb. However, many further measurements have confirmed that this noise has the spectrum of wavelengths expected to be radiated by a body when it is at an absolute temperature of 2.7 degrees Kelvin (minus 270.4 degrees Celsius), just slightly above absolute zero. In this case, however, a single body is not radiating, but rather the whole universe!

The observation was not unanticipated. A group at Princeton University were simultaneously looking for just this effect, as were a group of Soviet scientists. About ten years earlier George Gamow had proposed that the chemical elements were produced in the early minutes of the Big Bang and that a very low temperature remnant of photons should be left over.

Gamow, who was one of the great popularizers of science, coined the word *ylem* to describe the matter in that early moment of the Big Bang in which today's familiar matter was "cooked." Unfortunately, the theory ran into trouble when calculations indicated that the observed abundances of heavy elements were too great to have occurred in this way. Today we know that these are produced in supernova explosions, but that most of the helium in the universe was undoubtedly produced by the Big Bang. While the relative abundances of various elements compared to hydrogen varies from place to place in the universe, everywhere we look we see about the same ratio of one helium atom for every ten hydrogen atoms—strong evidence that they were mixed together in the Big Bang, which then spread them uniformly throughout the universe.

The Bell Labs scientists were smart as well as lucky. After the careful checking that necessarily accompanies any new discovery, they finally convinced themselves that the noise in their receiver was not being generated by the instrument itself. They published a paper with the sort of noncommittal title often found in scientific articles, "A Measurement of the Excess Antenna Temperature at 4080 Megacycles per Second." The interpretation of the data was given in an accompanying article by Dicke, Peebles, Roll, and Wilkinson. The Big Bang is now nothing but a quiet hiss, but that hiss had been heard.

6

The Fundamental Chaos

Natural science is not concerned with Nature itself, but with Nature as man describes and understands it. This does not mean that an element of subjectivity is introduced into natural science—no one claims that the processes and phenomena that take place in the world are dependent on our observations—but attention is brought to the fact that natural science stands between man and nature and that we cannot dispense with the aid of perceptual concepts or other concepts inherent in the nature of man.

Werner Heisenberg

Isaac Newton held many beliefs that were viewed as heretical in his era. At the same time that he was the Lucasian Professor of Mathematics at Trinity College, Cambridge University, he rejected the notion of the Trinity of God. But he was nonetheless a deist. Newton's greatest heresy was in rejecting the authority of scripture, though he was as well versed in it as anyone of his time. His authority was nature, and in uncovering the laws of nature he believed he was reading the mind of God.

All religions have their authorities, to which they refer for evidence for their particular interpretation of the "truth." This use of scripture as authority has its origins in those days when the symbols of writing were regarded as magical by most of humankind, their secrets jealously guarded by a small elite. It is difficult for us today to imagine how amazing it must have seemed to common people that a few scratches on a stone or a piece of parchment could carry the history of great events. Surely those events must have appeared supernatural and the people who participated in them looked superhuman in their eyes. Even the mundane records kept on the various economic activities of the tribe would have testified to the great power of the written word and its acceptance as a sacred authority, not to be questioned or doubted.

By Newton's day, knowledge of the written word was no longer restricted to the few and the printing press was making possible the wide dissemination of all sorts of ideas. Familiarity breeds comprehension, if not contempt, and the spread of the printed word carried with it the recognition that not everything that appeared in print was sacred, or even necessarily correct. Newton and the new "priesthood" preached the authority of observation over that of the printed word.

In the Newtonian view, the universe is a vast mechanism, a magnificent clock wound by the Creator and left by Him to tick away according to His perfect laws. In fact, an omniscient God never needs to intervene in nature since he would have anticipated every future event in that initial legislative process before time began. To Newton the bible was not God's scripture, but the universe itself. And he had been blessed to read the secrets of that scripture better than anyone before.

The concept of the mechanical universe implies that all events happen with complete predictability as they follow natural law. Each object in the universe is composed of atoms. If the initial position, mass, and velocity of an atom is known, then the laws of mechanics, set down by the Creator, enable us to calculate exactly what the future motion of that atom will be. That is, we can predict precisely where it will be at any later time.

If this can be done for one atom, it can be done for all the atoms in a body, and for each body in the universe, from a virus to a human to a galaxy. Of course, the practical ability to predict everything that happens in the world was never achieved by Newton or his successors, but proponents of the mechanical universe would explain that this is only because of computational inadequacy. Someday, with sufficiently powerful computers, it might be possible to compute the outcome of a flip of a coin or the behavior of a person. Whatever the current limitations of human abilities, events in the Newtonian world machine are seen to happen in ways determined by the laws of nature and the events that precede them. This view is in perfect consonance with the concept of fate: all that happens has been written.

Most people think this is the view held by science today. Newtonian mechanics has given us much of the technology that is so familiar to twentieth-century people, so it is natural that the philosophy associated with these successes should be widely disseminated. But early in this century, classical Newtonian mechanics was shown to be basically incorrect. In its place we now have *quantum mechanics,* which has as its most important tenet the *Heisenberg Uncertainty Principle.* This principle places basic limits on the predictability of future events that no amount of computer power will ever be able to overcome. At the atomic level, and by extension at all levels, there is a fundamental randomness and chaotic behavior in matter, which even makes it possible to create something out of nothing in a *quantum fluctuation.* Perhaps such a random unplanned quantum fluctuation produced the universe. In any event, the evidence has surely turned against the mechanical universe and the existence of fate.

In quantum mechanics, as in relativity, we find ourselves making statements that defy common sense. We have already seen how often common sense can be wrong; common sense and objective observation of the real world are not necessarily the same. And if we are to refute some of Newton's ideas—but by no means all—we must still call upon Newton's authority of observation to do so. Quantum mechanics is not a theory simply invented by someone unhappy with the Newtonian view; rather, it was required by facts about the real world of our senses.

Just as the nature of light led Einstein to his theories of relativity, phenomena involving light also led to the quantum theory. In the late nineteenth century the electromagnetic wave theory of light was successful in explaining all but a few of the facts known about the behavior of light. There were basically three phenomena that the wave theory could not reconcile: (1) the so-called *black body radiation* from a body; (2) the *photoelectric effect,* and (3) the *spectral lines* of atoms.

Every body radiates electromagnetic waves by virtue of the fact that there are charged particles in each body that are forever in motion. The sun and stars, or a heated coal on a fire, radiate visible light; but the bodies of everyday experience radiate mostly in the invisible *infrared* region of the spectrum, and we see them by their reflected light, rather than by what they radiate. Our eyes are insensitive to the long wavelengths that each of us radiate at one another, though we can be seen in the dark with an infrared camera or "snooperscope." A perfectly black body reflects no light, so the radiation from such a body is called *black body radiation* (fig. 6.1). Although the sun and heated coals are not black in our common usage of color, the term is still applied to these objects when they radiate with the type of wavelength spectrum that characterizes black bodies.

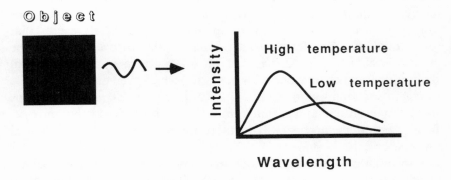

Fig. 6.1. Black body radiation is observed from every object. The intensity varies with wavelength as shown in the graph. The peak wavelength depends on the temperature of the body, with the intensity at a higher temperature peaking at a lower value of wavelength than that for a body of lower temperature. Classical wave theory predicts infinite intensity at zero wavelength, in disagreement with observation.

A simple model of a black body is an empty box, made of copper so as to conduct electricity, and painted black, with a tiny pinhole in one side. The electrons in the walls of the box are free to move about and do so simply under the action of collisions with atoms. These atoms move randomly about just from the thermal energy contained in the body. The electromagnetic waves set up are those whose wavelengths neatly fit inside the box, which acts as a *cavity resonator*. The problem is that the smaller the wavelength of the wave, the more waves can fit inside the box. So if we calculate the expected *spectrum* of the radiation from the pinhole in the box, we find increasingly more energy at shorter wavelengths.

This is called the *ultraviolet catastrophe,* since it occurs in that region of the spectrum for most bodies. Obviously bodies do not radiate infinite amounts of energy at short wavelengths. The wave theory has to be wrong; clearly there is something cutting off the shorter wavelengths that is not accounted for in the classical wave picture.

In 1900 Max Planck (1858-1947) introduced the idea of the *quantum* to explain this cutoff and describe the spectra of black bodies. Planck proposed that light is not emitted continuously but in individual packets, which he called *quanta*. The energy of these quanta is proportional to the frequency of the light, $E = hf$, where frequency is designated by f and h is a constant, now called *Planck's constant,* which had to be measured experimentally. Planck was able to explain the observed spectra of black bodies perfectly, getting exactly the right shape of these spectra for any given temperature of the body. He fit all the data for different temperatures with a single value of h ($h = 4.14 \times 10^{-15}$ electron-volt-seconds), where the electron-volt, abbreviated eV, is the standard unit of energy in atomic physics and h has the dimensions of energy multipled by time.

The spectra cut off at short wavelengths, which are the same as high frequencies, simply because of energy conservation: a body of a given temperature and a given number of particles contains only so much energy and can radiate only a limited number of the higher frequency electromagnetic waves, which carry higher energy.

The connection established by Planck related the energy carried by light to its color. The classical electromagnetic theory of light recognized that a light wave carried energy from one point in space to another, just like a sound wave or a sea wave; otherwise, these could not set particles in motion, which requires the input of energy. However, the classical theory viewed the phenomenon of color as something separate, associated with the wavelength rather than the energy. As with so many revolutionary concepts, Planck's hypothesis did not add to knowledge so much as unify it. Time and again in our story, we find that older ideas are replaced by a smaller number of independent new ideas. Newton unified the gravity of the heavens with gravity on earth. Maxwell unified electricity and magnetism. Einstein unified space and time, and matter and energy. Planck unified energy and color.

Max Planck was a cautious, careful revolutionary. While proposing that light is not continuous but discrete, he did not go so far as to suggest that light was corpuscular in nature. This was Newton's old idea, which had been discarded as a result of the success of Huygens's wave theory. It was not possible simply to return to the particle picture of light without running up against all the experimental evidence used to discard that picture in the first place.

Einstein was more daring. In that fantastic year 1905 when he proposed the Special Theory of Relativity and published the equations that enabled Avogadro's number to be determined from Brownian motion, Einstein proposed that light is indeed composed of particles, which he called *photons.*

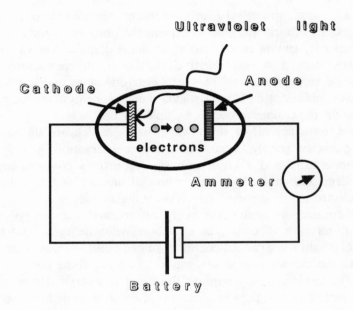

Fig. 6.2. In the photoelectric effect, ultraviolet light incident on the cathode of a vacuum tube causes electrons to flow from the cathode to the anode, producing a flow of current in the circuit, as measured by the ammeter.

Einstein's paper showed that the particulate nature of light is needed to explain the *photoelectric effect*. It had been observed earlier that ultraviolet light hitting the cathode of a vacuum tube could induce a current to flow even against a back voltage. This could be understood as occurring when the light kicked electrons out of the cathode metal with enough energy to overcome the back potential and reach the anode of the tube (fig. 6.2). The puzzle, however, was that for the cathode materials generally in use, visible light—no matter how intense—could not generate more than the feeblest current, while even the lowest intensity ultraviolet light gave a hefty current.

If light were a stream of waves, then increasing the intensity of the light should increase its energy to the point where it would be sufficient to knock electrons out of the cathode. Certainly a pebble resting in a crevice of a larger rock on a beach will be dislodged by a powerful ocean wave. We would not expect tiny high frequency ripples to do what a ten-foot monster wave could not.

Yet this is what happens in the photoelectric effect, and Einstein explained it as follows: higher frequency ultraviolet light is composed of photons, which by Planck's relation $(E = hf)$ have higher energy than the photons of visible light. When an individual photon strikes an electron bound in the metal, it can transfer its energy to that electron. If that energy is just sufficient to remove the electron from the metal and get it over to the anode, then a current will flow. Raising the intensity of the light does not help, because this only increases the number of photons striking the metal. It does not change the energy of the individual photons; only raising the frequency can do that.

Einstein then predicted that the frequency at which current will begin to flow will be proportional to the back voltage applied between the anode and cathode of the vacuum tube, and that the proportion will be determined by Planck's constant, h. After several years of careful experimentation, in 1916 Robert Millikan (1868–1953) published results in which Einstein's hypothesis was confirmed and Planck's constant measured to a precision of 0.5 percent. The value agreed precisely with that inferred by Planck from black body spectra. Thus Planck's idea was transformed from a model, which gives a nice mathematical description of experimental data, to a fundamental concept that can be used to explain different and seemingly unrelated phenomena.

Since then individual photons have been counted in numerous experiments using extremely sensitive light detectors call *photomultiplier tubes,* in which the electrons generated when light hits a cathode are multiplied many times by being accelerated by high voltage down a chain of electrodes. Cathode materials have been developed that emit electrons when struck by visible light. These and other sensitive instruments allow the effects of individual photons to be studied in great detail.

The interaction of individual photons with matter has been observed in many contexts. For example, in *Compton scattering* an X-ray photon scattering from an electron loses energy and thus increases in wavelength (fig. 6.3). The wave theory of light is incapable of explaining this observation. Today the evidence for the particle nature of light is overwhelming. The electromagnetic wave theory, despite its great success in explaining and predicting the behavior of light, is nonetheless wrong in some contexts where it disagrees with experiment.

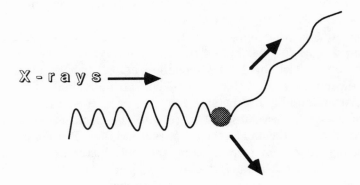

Fig. 6.3. In the Compton effect, X-rays scattered by electrons in a crystal experience an increase in wavelength.

If the wave theory of light fails to explain these experimental facts, how do we then reconcile the interference and diffraction effects that led physicists to discard Newton's corpuscular theory and adopt the wave theory in the first place? In 1921 a French aristocrat, Louis de Broglie (1892–1987), was refused participation in the third of the Solvay conferences, which had become the meeting places of the great physicists who were at that time revolutionizing the world of science. De Broglie was not yet known, and even princes do not automatically qualify for participation in the rites of the new priesthood. By 1927, however, de Broglie had joined those illustrious ranks by a great feat of intuition, which solved the wave-particle paradox.

De Broglie turned the question around. Instead of asking why photons have wave-like properties, he suggested that not only photons but all particles have wave-like properties. He viewed this as a fundamental part of the description of their behavior. De Broglie proposed that particles and waves are not separate phenomena, as had been previously thought; thus, yet another unification of concepts was added to our ever-growing list.

De Broglie made this quantitative by proposing a relationship between the wavelength (λ) of a particle, which determines its wave properties, and its momentum, p, which determines its particle properties. He assumed that this relationship was the same as that for photons: $\lambda = h/p$. This can be demonstrated as follows. As discussed earlier, momentum was first proposed by Newton as the quantity of motion of a body, since it carries information about both the inertial mass m and velocity v of the body. Specifically, $p = mv$. We have seen that photons also have inertial mass according to Einstein's relation $E = mc^2$. Since $v = c$ for photons, the energy of a photon and its momentum are then simply related by $E = pc$. Planck said that the photon's energy is proportional to frequency by $E = hf$. So we have, from Einstein and Planck, $pc = hf$. Making one final substitution of the relation between the frequency and wavelength of a wave. $f = c/\lambda$, we obtain $\lambda = h/p$.

We have derived this result for photons; De Broglie simply proposed that this relation holds not just for photons but for all particles, including electrons.

In 1924 de Broglie submitted a dissertation on his "matter waves" to the Sorbonne for a Ph.D. degree. The faculty did not know quite what to do with it, so unconventional were the ideas, but his paper began to be distributed and was read by Einstein and others. It soon became recognized as another important step in the fundamental understanding of the stuff that constitutes our universe.

De Broglie said that particles are waves and waves are particles. Sometimes this is called the *wave-particle duality,* but I think this term is misleading. Rather, it should be called the *wave-particle unity.* Light, thought to be waves, has particle-like properties as evidenced by the photoelectric effect and the other experiments discussed earlier. Now we find that electrons, neutrons, and even macroscopic bodies should exhibit the same sort of diffraction effects as light, bending around corners and otherwise behaving like waves.

Electrons had been observed to diffract as early as 1921 by scientists at Bell Labs; however, the definitive experiment was reported in 1927, after de Broglie's proposal. This experiment verified that electrons do indeed behave as waves, with a measured wavelength, h/p, as predicted.

While all this was going on, another part of quantum mechanics was being developed by Niels Bohr (1885-1962). The matter wave idea would make it all begin to fit together. We must backtrack a bit to pick up this development.

In 1885 a Swiss high school teacher, Johann Jakob Balmer (1825–1898), had noticed regularities in the frequencies of the spectral lines of atomic hydrogen. We have already noted how atoms often radiate with very precisely defined colors. These are the spectral lines Balmer studied. He found that the visible spectral frequencies obeyed a formula,

$f = R(1/4 - 1/n^2)$, where R is a constant (the Rydberg constant), and n is a positive integer greater than two.

Bohr was able to derive the Balmer formula using the model of the atom proposed in 1911 by Ernest Rutherford (1871-1937). In our discussion so far, we have been using the term *atom* in the ancient Greek sense, to refer to the elementary particles of nature. We have seen that the chemical elements are particulate, and that these specific particles are what we continue to call *atoms* though we now know that they are not elementary. By the early 1900s physicists had recognized that atoms had a substructure of their own. J. J. Thomson (1856–1940), the discoverer of the electron, had imagined the atom to be composed of a positively charged fluid embedded with negatively charged electrons, the so-called *raisin pudding model*. Others had considered a planetary picture in which the electrons orbited like planets in the solar system. This *Saturnian model,* however, was found to be unstable and initially discarded.

Experiments in which the alpha particles emitted by radioactive atoms were scattered from gold foil convinced Rutherford that the data could not be explained by the Thomson model. The large scattering angles observed for the alpha particles could only come about if they in fact scattered from a tiny positively charged nucleus in the atom that contained most of the mass of the atom. Thus Rutherford in 1911 resurrected the Saturnian model of the atom: just as the planets circle the sun in a solar system that is mostly empty space, so too the electrons in an atom circle about a nucleus that is much smaller than the atom as a whole.

The idea that nature repeats itself on various scales is one that has captured the public imagination. Perhaps there are little solar systems inside the nucleus, and perhaps our solar system is an atom in a larger body, and so on. This is a great oversimplification, however, and the atom is really vastly different from the solar system in many fundamental ways. For example, the solar system is held together by gravity, while the atom is held together by electricity. These forces each have the same property of decreasing with the square of the distance, which explains the crude similarity of the two systems, but this is about all they share. The gravitational attraction between two bodies depends on their masses and is appreciable only when those bodies are of astronomical dimensions. It is negligible for the bodies of our normal experience, and even more so for the bodies inside an atom. The force that holds the atom together depends on the electrical charge of the bodies and, unlike gravity, is repulsive or attractive depending on the relative sign of the charges, positive or negative: opposite charges attract, while like charges repel. The orbiting electrons are negatively charged and are held in by their attraction to the positively charged nucleus. This entails a major problem, which does not occur for the solar system.

In classical electromagnetic theory, any accelerated charge will radiate electromagnetic waves. The electrons in orbit in an atom are accelerated toward the nucleus by the centripetal force of electrical attraction, just as any body moving in a circle is accelerated toward the center of that circle. Another way to see this is to imagine you are looking at the electron's orbit edge-on. You would see an oscillating charge, just like a standard dipole antenna, that would radiate an electromagnetic wave. So classical physics predicted that the planetary atom could not last very long, as the revolving electrons radiated away energy and spiraled into the nucleus. Yet this does not happen, and the planetary picture is consistent with observations. Once again scientists were forced to conclude that their theories were wrong because they failed to agree with experimental data. The theory that had to be discarded in this case, however, was no minor speculation in the literature but the whole framework of Newtonian physics, which had proven so successful for centuries. Classical physics was so beautiful and so successful that any Platonic philosopher would have concluded that it had to be right and the experiments wrong. But by the twentieth century, scientists had become fully prepared to discard any theory, no matter how beautiful and useful, if it failed to describe observations correctly.

Classical physics did not have to be thrown on the scrap heap by these results. Rather, it would be replaced by something more fundamental that would approach classical physics in the limit where the latter in fact does agree with experiment. Bohr recognized this, and called it the *Correspondence Principle*. Classical physics can certainly still be used and is used today to build bridges and machines and to compute the orbits of comets and space vehicles. We must be careful, however, not to apply it to situations where it gives the wrong results. Most, but not all, of these situations occur on the submicroscopic scale of atoms and nuclei.

Like most of the giants of early twentieth-century physics, Bohr was still a young man in his twenties when he made his unique contribution to that revolutionary era. People had realized that atomic spectra and the Balmer formula must have something to do with Planck's constant, *h,* which was called in those days the *quantum of action.* The problem was how to incorporate it, and most attempts were pretty muddled. It took an intuition in Bohr that Einstein later called a miracle, "the highest form of musicality in the sphere of thought," to infer the connection. Einstein said that he had had ideas along the same line but that most daring of all thinkers did not dare to pursue these fantastic thoughts.

Bohr made two seemingly contradictory assumptions: (1) the stationary, or equilibrium, states of an atom are composed of certain *preferred* electron orbits, and not all possible orbits are allowed; (2) these can otherwise be described by classical physics. The emission of light, he said, cannot be explained classically and occurs when an electron jumps from one of the preferred orbits to another. The frequency of that light is then, following Planck, $E/h,$ where E is the difference in the energy of the two orbits.

Using these assumptions, Bohr was able to derive Balmer's formula for the spectral lines of hydrogen, including the correct value of the Rydberg constant, R. This result was first published in 1913 when Bohr was 28. At the very end of the article he stated the condition for the allowed orbits: The angular momentum is an integral multiple of $h/2\pi$. This *quantization condition* appeared rather obscure at the time, but no one could deny that it worked. Later, after de Broglie had introduced his idea of matter waves, Bohr's condition began to make a certain amount of sense. His "allowed orbits" turned out to be just those in which an integral number of the de Broglie wavelengths fit around the circumference of the orbit. In other words, the stationary states of the atom were seen to be a sort of resonance set up inside—musical notes like those of a plucked violin string. Perhaps that was what Einstein was thinking when he referred to the "musicality" of Bohr's hypotheses.

One of the objections that Rutherford expressed toward the Bohr atom was that the electron somehow had to know beforehand where to jump. Up to this time all physics was causal—events were seen to happen because something caused them to happen. In the Bohr atom, however, electrons just jumped from one energy level to another, with no explanation as to why they jumped or how we would go about predicting when a jump would take place.

In 1917 Einstein suggested that causality may not strictly apply in the atomic domain. Instead, atomic transitions occur unpredictably: we cannot predict whether a particular atom will emit light, but given a large number of such atoms, as usually exists in any practical experiment, we can predict the fraction of them that will emit light in a given time interval.

This was the first glimpse of what was to become the most revolutionary aspect of quantum mechanics—its fundamentally statistical and nondeterministic nature. This idea will allow us to conclude eventually that the creation of the universe itself was a random event, without cause or plan. The theory of the quantum developed in the 1920s by Werner Heisenberg, Erwin Schrödinger, Wolfgang Pauli, Max Born, Paul Dirac, and others showed how to calculate the probabilities for atomic events to happen. But for the most basic problem of all—the prediction of the motion of bodies—the great founders of quantum mechanics found that the probability could never be 100 percent and that, at the most fundamental level, particles behaved in a way that was not determined by any known *or knowable* laws of nature.

Modern quantum mechanics, which replaced the "old quantum theory" of Planck and Bohr, began around 1924 with Werner Heisenberg (1901–1976) and Erwin Schrödinger (1887–1961) working independently along different lines. Usually physics students learn Schrödinger's *wave mechanics* first, with its slightly less advanced mathematical structure and its somewhat intuitive representation of de Broglie's matter waves by the *wave function*. However, Heisenberg's *matrix mechanics* is considerably more general and also was published first. Schrödinger did not like the "transcendental algebra" of matrix mechanics and tried to give the theory a more visual reality, which Heisenberg, in turn, called "trash." Heisenberg insisted that the only useful thing that Schrödinger did was to show how to calculate *matrix elements,* and he was able to prove that the Schrödinger wave function is a purely mathematical artifact and not the physical entity that Schrödinger claimed. Ultimately, Paul Dirac (1902-1984) showed that both forms of quantum mechanics are equivalent. Today, most physicists do not fret about the fact that quantum phenomena cannot be readily visualized in terms of the objects of common sensory perception. After all, quantum phenomena belong to a world inaccessible to our normal senses.

Since ancient times, numbers have been used to represent quantitative measures of the observable phenomena of the real world, from the height of a pyramid to the weight of a basket of grain. Numbers, however, are but one class of mathematical objects—those which obey the well-known rules of arithmetic. Heisenberg discovered that quantum mechanical observables are not best represented by numbers but by other mathematical objects, which do not obey all the rules of conventional arithmetic. In particular, he found that the *commutative law,* which says that $AB = BA$, is not necessarily obeyed by two quantum mechanical observables, A and B. Tables of numbers called *matrices* are one particular set of mathematical objects that are not commutative, and Heisenberg used these to represent observables.

In Schrödinger's wave mechanics, observables are represented by the differential operators of the ordinary calculus taught to college students; these also are not commutative. Dirac recognized that the noncommuting property was more fundamental than either of the specific representations—differential operator or matrix—and that the observables were more generally operators in an abstract *Hilbert space,* where the vectors in the space represent the various quantum mechanical states of the system. This is the approach used today in calculations of the quantitative behavior of atomic and subatomic processes.

Although quantum mechanics essentially only enables us to compute probabilities, its applications are wide-ranging and enormously practical. The detailed structure of the hydrogen atom can be calculated to many significant figures of accuracy, far beyond what is possible in the crude Bohr model. Quantum mechanics fully explains the fundamental principles of chemistry and the Periodic Table of the elements. Virtually no mystery remains today about the behavior of matter as it appears to us in the solids, liquids, and gases of our everyday experience; if something is not calculable, then it is probably random. The remaining mysteries of a fundamental nature concerning matter involve behavior at the smallest distances, much less than the size of an atomic nucleus, and at energies or temperatures far above those which exist in normal matter on earth. During the last half-century this knowledge has been exploited in the remarkable synthetic materials and solid-state devices, such as computer chips, of today's technology.

Just as the success and tremendous applicability of Newtonian mechanics was the best evidence for its validity, so too the success of quantum mechanics convinces us that it must contain many elements of truth. Today the apparent contradictions in Bohr's hypotheses, and the general unease felt about Planck's law and de Broglie's matter waves, have been satisfactorily resolved by the full theory of quantum mechanics. Quantum mechanics has been placed on a logically sound and mathematically precise footing.

Any fundamental theory must begin with a set of assumptions, or *axioms,* that cannot be derived from anything more fundamental but must be simply stated as given. The ultimate validity of these axioms necessarily must rest on the success of the theory derived from the axioms —its agreement with observations about the real world and its ability to predict previously unrecorded observations.

Since axioms are human inventions, it should be no surprise that different axioms can be assumed, which lead to the same conclusions. We choose between these on the basis of utility or simplicity. One way to derive most of quantum mechanics, including the apparent wave nature of particles, is to take as a fundamental axiom the *Heisenberg Uncertainty Principle: It is impossible to measure both the position and momentum of a particle simultaneously with infinite precision.*

Certainly Heisenberg did not simply "make up" the Uncertainty Principle. Just as the axioms of Euclid were undoubtedly inferred from geometrical figures, the Uncertainty Principle was not the result of Kantian "pure thought" but was based on the development of quantum mechanics up to that point. Heisenberg went back and asked a basic question: What is actually done when we make an observation or measurement?

The only way we can observe and measure something is to interact with it in some way. The act of observation, then, implies disturbing the system being observed. When we look at something we see photons bounce off it, either from the sun or artificial light. In principle these photons hitting the object will cause it to recoil backward (fig. 6.4). That is what would happen if, for example, we used machine gun bullets instead of photons to look at another person. Photons carry comparatively little momentum, however, and we do not even notice the slight recoil when we shine a flashlight at someone in the dark.

Heisenberg noted, however, that the situation can be quite different when we shine light on an atom, or an electron inside an atom. Perhaps the most basic act of measurement is to measure the position of a particle. Suppose we wish to measure the position of an electron in an atom

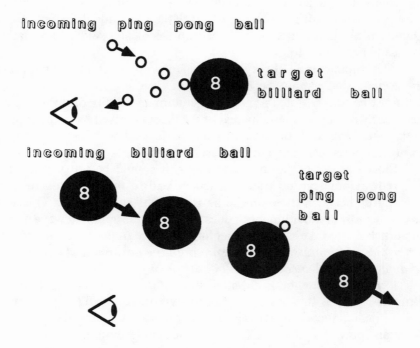

Fig. 6.4. In order to "see" an object with particles, the object must have sufficient mass to scatter the incoming particles. Here ping pong balls are easily scattered by a billiard ball into the detector, symbolized by the eye. Incoming billiard balls, however, are not scattered by a ping pong ball and so are not "seen."

by shining light on the electron and then recording the light that bounces off. If that light is viewed as a wave, the wavelength of that wave must be small compared to the uncertainty in position that we wish to determine (fig. 6.5).

This is no different from the case of looking at microbes under a microscope. Since visible light has a wavelength of about one micron (10^{-6} meter), we cannot really see microbes less than about 10 microns in size. To do that we must resort to X-ray or electron microscopes in which the wavelength is smaller. An atom is about 10,000 times smaller than the wavelength of visible light, so we can never hope to see an

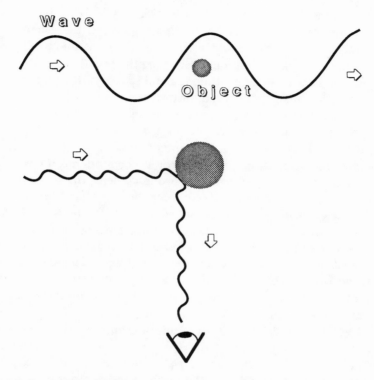

Fig. 6.5. In order to "see" an object with waves, the wavelength of the waves must be smaller than the size of the object. In the top figure, the object is smaller than the wavelength and the wave is undeflected. In the bottom figure, the wave is deflected into the detector, symbolized by the eye.

atom, or the electrons within, with visible light no matter how great we make the magnification of our microscopes. The atom is forever inaccessible to our eyes, but not to our brains, which can interpret the data from the instruments we build to aid our eyes. Let us still speak of "seeing" when we use these instruments.

To "see" an electron in an atom we would need to hit it with "light" having a wavelength of about 10^{-5} micron (10^{-11} meter, one tenth the size of the atom). From de Broglie's relationship, this photon will have a momentum, $p = h/10^{-11} = 4 \times 10^{-4}$ *eV-seconds/meter*. All this momentum can be transferred to the electron, in which case it obtains a velocity, $v = p/m = pc^2/mc^2 = 7 \times 10^7$ *meters per second,* 23 percent of the speed of light. (For an electron, $mc^2 = 5.11 \times 10^5 \ eV$, and recall that $c = 3 \times 10^8$ *meters per second.*) If that happened, the electron would not remain in its orbit. In other words, it is impossible to look at the electron inside an atom without destroying the atom!

This is, in a nutshell, the fundamental source of the unpredictability of atomic phenomena. If we cannot measure the position of an electron in an atom, *it does not have such a position.* Time is what you measure on a clock. Position is what you measure by bouncing light off the object onto a meter stick. If you cannot observe something, then that something has no meaning. The electron in the atom has no well-defined position; all we can say is that it is "probably here" or "probably there." Quantum theory tells us how to compute those probabilities.

The electron has some probability of being found far from the nucleus of its atom, or even in another nearby atom. This results in an attractive force between atoms, which is as important in binding them together into molecules as are the classical electrical forces. The existence of this *exchange force,* which has no classical analog, can be regarded as another experimental verification of the validity of the non-common sense ideas of quantum mechanics.

Beyond the understanding of atoms and molecules, the Uncertainty Principle forces us to discard the essence of Newtonian mechanics. In Newtonian mechanics, the position and momentum of a body are the two complementary variables that describe a body's motion. It was always assumed that these could be measured to any desired precision at any time. The Heisenberg Uncertainty Principle says that the product of the uncertainty in the position and the uncertainty in momentum must always be greater than Planck's constant, h. To measure the position more accurately you must reduce the wavelength of the light or other probe being used in the measurement. This raises its momentum, making the momentum of the body being observed increasingly uncertain, since that momentum can be transferred when the probing particle collides with the body.

This means that the motion of bodies can never be predicted exactly. *There is a fundamental unpredictability to nature.* Things may happen with a certain orderliness, and even a certain predictability, but the certainty is never 100 percent. And there are some phenomena that happen spontaneously: an electron jumps from one energy level to another in an atom; an atomic nucleus decays. Not everything that happens has a cause. The universe, after all, is not a vast Newtonian machine with everything determined by initial conditions set down at the creation. All things do not happen because it is written; some things just happen.

On the vast scales of astronomical bodies in the universe, and even on the more modest scales of everyday experiences, this fundamental chaos is not always evident. As we go to smaller and smaller distances, however, our deterministic prejudices increasingly collide with reality. Our discussion will lead us back to a time when the whole universe was far smaller than the nucleus of an atom and where the Uncertainty Principle makes it impossible to even define such common sense notions as space and time or particles and waves. At that time, if we can still call it a time, causality and determinism also cease to have any meaning.

These were the conditions under which the universe began to form into what we see it to be today. When we speak of that time, we will have to be careful to avoid terms that really do not apply because they cannot be logically defined. One such term is *creation,* which implies a causal relationship that is meaningless at very small distances. If the universe happened in a flash of light, it just happened, as an electron in an atom just happens to jump orbits. In that initial quantum fluctuation many universes were probably generated, with ours and its special forms and structure representing just one particular roll of the dice.

7

Naked Bottom and Other Quarks

Der kleine Gott der welt bleibt stets von gleichem Schlag
Und ist so wunderlich als wie am ersten Tag.
Ein wenig besser würd er leben,
Hätt'st du ihm nicht den Schein des Himmelslichts gegeben;
Er nennt's Vernunft und braucht's allein,
Nur tierischer als jedes Tier zu sein.
Er scheint mir, mit Verlaub von Euer Gnaden,
Wie eine der langbeinigen Zikaden,
Die immer fliegt und fliegend springt
Und gleich im Gras ihr altes Liedchen singt.
Und läg er nur noch immer in dem Grasel!
In jeden Quark begräbt er seine Nase.*

Goethe, *Faust*

We and our universe are the result of our atoms, and these atoms are
the result of their parts (fig. 7.1). Thus, if we are to understand how
the universe came about, we must learn about the parts of atoms, the
elementary particles. The last decade has witnessed the coming together

---

* The small god of the world will never change his ways
And is as whimsical—as on the first of days.
His life might be a bit more fun,
Had you not given him that spark of heaven's sun;
He calls in reason and employs it, resolute
To be more brutish than is any brute.
He seems to me, if you don't mind, Your Grace,
Like a cicada of the long-legged race,
That always flies, and, flying, springs,
And in the grass the same old ditty sings;
If only it were grass he could repose in!
There is no trash he will not poke his nose in.

(translation by Walter Kaufmann)

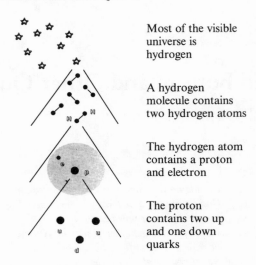

Most of the visible
universe is
hydrogen

A hydrogen
molecule contains
two hydrogen atoms

The hydrogen atom
contains a proton
and electron

The proton
contains two up
and one down
quarks

Fig. 7.1. The layers of matter.

of two previously separate disciplines—elementary particle physics and cosmology. Cosmologists have learned that the physics of the early universe is just that being studied by particle physicists with giant particle accelerators, and particle physicists have been surprised to find a profound application for their efforts.

When two protons smash against one another in a colliding beam accelerator, the energies involved correspond to temperatures that existed one trillionth of a second after the universe began. In this chapter and the next, we will explore what has been learned from accelerators that makes it possible, for the first time in history, to visualize the conditions existing shortly after time began. The story begins with atoms.

What we now call an atom is a positively charged nucleus surrounded by a cloud of negatively charged electrons, which because of the Uncertainty Principle are not in clearly defined planetary orbits (fig. 7.2). This fuzzy picture of the atom is unsatisfactory to most people when they hear about it for the first time.

We all seem to grow up needing concrete models based on our own experience in order to convince ourselves that we understand something, and it is a great intellectual feat to be able to cast the prejudice aside. For example, physics students eventually become quite comfortable with the quantum picture of the atom, as they learn that the human

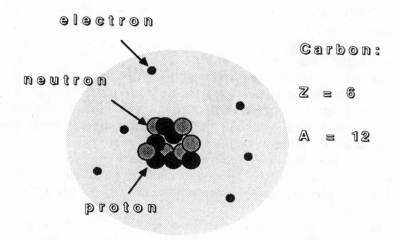

Fig. 7.2. An example of the nuclear atom. Shown is a carbon atom which contains $Z = 6$ protons and $A - Z = 6$ neutrons in the nucleus, where in general Z is the atomic number and A is the atomic weight. When the carbon atom is electrically neutral, this nucleus is surrounded by a cloud of $Z = 6$ electrons.

mind is not limited to grasping visual concepts. Ideas expressed in terms of abstract symbols and mathematical equations can be as real as those visualized in terms of the objects we see around us. These, after all, are also abstract images within our brains.

After students of quantum mechanics have mastered the ability to solve its equations and apply them to real problems in chemistry or physics, they see how well the process works in getting results that agree with experiment. At that point, after years of experience, they convince themselves that they understood it as soon as they became capable of making those calculations. You do not understand a concept until you are able to apply it, and you can be sure you do understand it when you can apply it.

It still is fruitful to continue to think of the atom as being composed of fundamental particles, which we visualize as tiny points—as long as we recognize that this is just imagery. We cannot look inside the atom and see particles, but when we hit an atom with enough energy to destroy it, particles come out and can be seen, or at least unambiguously

identified by instruments. Further, this picture of the atom works—
it fully explains and predicts the behavior of the matter composed of
these atoms.

Among the most fundamental scientific questions one can ask are:
What are the elementary constituents of the universe? How do they
interact to produce the phenomena we see? Contemporary physics has
reached a remarkable level in its preliminary answers to these questions.
The current view, by no means final or complete, has successfully reduced
all the observed world to a comparatively small number of fundamental
objects, acted on by three or four forces that may be unified into one
force before the end of this century. We will now see how all this came
about in the remarkably short period of a half-century after the discovery
of the structure of the atom and the quantum mechanics which describes
that structure.

Before picking up that story, however, let us briefly place it in
the perspective of the much longer search for the elementary constituents
of matter, which began with the ancient Greeks. The general consensus,
which was developed by the Greeks and carried into the Middle Ages,
was that there are four elements of matter: water, earth, air, and fire.
Most bodies were believed to be a mixture of these in various proportions.
Gold, for example, was said to be shiny because it contained more
fire than the typical chunk of rock lying on the ground. While this
idea crudely explained many of the qualities of matter, it ultimately
was discarded because it simply did not work. Alchemists for centuries
had tried to manufacture gold by adding fire to various materials,
particularly lead, but they always failed. Even Isaac Newton spent many
years in fruitless alchemical experimentation.

Eventually, alchemy developed into the science of chemistry, culmi-
nating in 1869 with the discovery of the Periodic Table of the chemical elements by a Russian, Dmitri Mendeleev (1834–1907), following
a slightly earlier, less complete version by John Newlands in England.
Mendeleev's table succeeded because it worked. It listed elements by
their properties and showed that they followed progressions in these
properties. Elements were later found where gaps existed in the table,
and which possessed the properties predicted by that particular position
in the sequence. The element germanium, very important in today's
technology, is one example of an element found after it was predicted
by the Periodic Table. Chemists called these substances *elements* because

they could not break any of them down to more elementary constituents, and could use them to build known—and, eventually, previously unknown—compounds. Contrary to alchemists' hopes, however, gold could not be manufactured—because it is an element.

We have already seen that early in this century matter was finally shown to be fundamentally particulate. The chemical elements were associated with atoms, and atoms were soon found to have a substructure of nuclei and electrons. The discovery of the atomic nucleus had been primarily the result of the observation of the natural radioactivity of certain atoms. Recall that experiments with alpha rays, one form of natural radioactivity, led to Rutherford's proposal of the nuclear atom. It was recognized that an atom undergoing radioactive decay actually changes from one chemical element to another. The alpha particle emitted in alpha radiation is in fact a nucleus of helium, the second element in the Periodic Table after hydrogen. When the nucleus of an element emits an alpha particle it is transmuted into a new element two units lower in the Periodic Table.

In radioactive beta decay, an electron is emitted from the nucleus, and since the electron carries away a unit of negative electric charge, the resulting nucleus moves up one unit in the Periodic Table.

These types of transmutation of the elements, including that of lead into gold, have become commonplace. The dream of instant riches, however, has not been achieved by modern day alchemists since the value of the small bits of gold produced is far less than the cost of the process that produces them. What alchemists and chemists failed to achieve is now a regular occurrence in nuclear physics labs. There is nothing mystical about it.

The third form of nuclear radioactivity is the emission of gamma rays, which are protons of very high energy or short wavelength, even shorter than X-rays. That is, they are a form of light in a region far beyond the visible part of the electromagnetic spectrum. Since photons carry no electric charge, a nucleus emitting a gamma ray goes from an excited energy state to one of lesser excitation without the atom containing the nucleus changing from one chemical element to another. Gamma ray emission on the nuclear scale is what ordinary photon emission is at the atomic and molecular scales.

The elements, or atoms, are not elementary—they have more fundamental parts, which can be rearranged to produce other varieties of atoms. The key lies in the nucleus and its structure. The simplest atom, hydrogen, is composed of a single electron orbiting about a nucleus, carrying one unit of positive electric charge. This nucleus is called a *proton*. To most physicists in the early 1930s, it seemed unlikely that the proton and the ninety-one other nuclei of the chemical elements were separate fundamental particles. Probably, they also were composite bodies and the only question was the nature of the components.

One idea that occurred to many was that the nucleus is composed of protons and electrons; the fact that some nuclei emit electrons in beta decay is certainly an indication that electrons are inside. There is also an appealing symmetry and elegant simplicity to a world made up of one particle of positive and one particle of negative electricity. But for a number of reasons, mainly incorrect predictions of the spins of nuclei, this did not seem to work. Then, in 1932, the *neutron* was discovered and established as the other nuclear constituent. The story of this exceptional year in science is a fascinating one.

One of the finest scientists of the century was Marie Curie (1867–1934), the first person ever to win two Nobel prizes. One of these was shared with her husband Pierre in 1903 for their work on radioactivity. Pierre was tragically killed by a carriage on the streets of Paris in 1906. Their daughter Irène had been reared to follow in her illustrious parents' footsteps, and in 1931 was working with her husband, Frédéric Joliot on the extremely penetrating radiation from beryllium, which in turn had been activated by radioactive polonium. The Joliots found that this radiation was able to eject protons from paraffin, a result which they confirmed by photographing the proton's track in a cloud chamber.

The experiment was brilliant, and if they had interpreted it correctly, Irène would have duplicated the feat of her parents in sharing a Nobel Prize with her spouse for a very important discovery. Unfortunately the Joliots were a bit too conservative and misinterpreted the radiation as being gamma rays that kicked out protons from the paraffin in a nuclear version of the Compton effect. When their paper arrived in England, a young scientist named James Chadwick (1891–1974) took it to Rutherford, who quickly exclaimed, "I do not believe it." Calculations indicated that it was extremely unlikely that gamma rays could kick protons out of atoms the way X-ray photons can kick out electrons in the Compton effect.

An Italian physicist, Ettore Majorana, reading the same paper in

Rome said, "What fools, they have discovered the neutral proton and do not recognize it." In the space of one month Chadwick repeated and improved the Joliots' experiment, and sent his results to the journal *Nature* with the correct interpretation: the radiation from beryllium was composed of neutral particles with a mass almost the same as the proton. He called this new elementary particle the *neutron*.

Chadwick's discovery of the neutron had immediate application in understanding the slightly earlier observation by Harold Urey, F. G. Brickewedde, and G. M. Murphy of heavy hydrogen, or *deuterium*. What they observed was hydrogen, as far as a chemist was concerned, but with twice the atomic mass of hydrogen. It was recognized that the deuterium nucleus is a bound system of a proton and neutron.

The neutron is a particle having only slightly higher mass than the proton, but no electrical charge. (Heavy particles such as the proton and neutron are called *baryons*.) Atomic nuclei are composed of protons and neutrons bound together by some previously unknown force; since the neutron is neutral this force cannot be identical with the electric force that holds the electron in the atom. Neither could this force be gravity, which is about forty orders of magnitude too weak. Deuterium contains a single proton and neutron forming a nucleus of what chemically is still hydrogen because the atom still contains only one electron. This was another example of the earlier discovery by Rutherford and Soddy that the nuclei of the chemical elements have different forms, or *isotopes*. This discovery was to prove important in the development of the nuclear bomb a decade later.

The neutron and deuterium were not the only great discoveries of 1932. This was also the year of the discovery of *antimatter*. In 1928 Paul A. M. Dirac had succeeded in finding a relativistic description of the electron, which implied the existence of what seemed to be negative energy states. At first he thought these might correspond to the proton, but then he was shown that the particle which occupies the negative energy states must have the same mass as the electron but opposite charge: an *anti-electron*. In 1932 Carl Anderson and Seth Neddermeyer were able to photograph the paths of particles in a cloud chamber and, by means of an externally applied magnetic field that curved these paths in different directions for different charges, demonstrated that they had observed positive electrons, or *positrons*. This mirror-image matter, or antimatter, was discovered in the same year as the neutron.

Dirac's equation not only predicted the existence of the positron, but also another very profound property of the electron—its *spin*, or intrinsic angular momentum. Spin had been observed experimentally but was not understood. Dirac showed that spin naturally followed as a consequence, when quantum mechanics and relativity were combined.

Anomalies in the spectral lines of atoms had led two young Dutch physicists, George Uhlenbeck and Sam Goudsmit, to propose in 1925 that the electron had a spin. They wrote a paper and gave it to their professor at Leyden, Paul Ehrenfest. He liked it and told them to discuss it with Hendrik Lorentz (1853–1928), one of the pioneers of relativity and a famous senior professor. Lorentz told them it was impossible since the speed on the surface of the electron would be greater than the speed of light, so they rushed back to tell Ehrenfest to withdraw the paper. But it was too late; he had already submitted it. Fortunately for the two young men, they had a superior who was willing to stick their necks out for them, for their paper proved correct.

Any rotating body has angular momentum. For example, it is the conservation of angular momentum that keeps the earth spinning on its axis and accounts for the elliptical shape of planetary orbits. But the electron in relativistic quantum mechanics is a geometrical point, with no detectable structure—yet it carries angular momentum. This is not the only mystery. Bohr's original quantization condition held that the angular momentum of an electron's orbit in an atom occurs in units of $h/2\pi$. Later it was shown that this is a general property of angular momentum in quantum mechanics. The electron spin, however, had a value of only half the fundamental unit, which is also very strange. Somehow the electron has some inner dimensions that manifest themselves in this way. Ultimately other inner dimensions of the elementary particles were discovered, and we are just today beginning to glimpse what they really mean.

So, at the beginning of 1932, there were ninety-four elementary particles: the photon, the electron, and the nuclei of the ninety-two chemical elements then known. With the discovery of the neutron, this number dropped precipitously to just four, with the proton and neutron seen as the basic constituents of nuclei. The proton and neutron were each found to have the same half-unit spin as the electron, which also explained how the observed spins of nuclei were sometimes unit and sometimes half-unit. The photon has one unit of spin.

In a matter of months, this remarkably simple picture was complicated by the discovery of antimatter, and in a few short years it was almost reduced to shambles by the observation of many new particles that could not be built up from photons, electrons, protons, and neutrons.

The first series of these new particles were observed, for the most part, as tracks in cloud chambers exposed to cosmic rays. Initially little theoretical guidance existed for the observations, but gradually a symbiosis between theory and experiment began to develop. This was to become a hallmark of the field of elementary particle physics after World War II. Although the discoverers claim that theory played no role in their experiment, the positron had been predicted by Dirac's relativistic quantum mechanics.

The positron was a major triumph for relativity combined with quantum mechanics. Eventually, relativistic quantum mechanics developed into *quantum field theory*, pioneered around 1948 by Richard Feynman, Julian Schwinger, Sin-itiro Tomanaga, Freeman Dyson, and others. (Amusingly, *Tomanaga* and *Schwinger* each mean *oscillator* in the languages of their origin, and this is what the quantum fields do— oscillate.) Quantum field theory succeded in describing the electromagnetic interaction between particles in a way that bore little resemblance to the methods of the previous century. Instead of an electromagnetic field acting on a charged particle to pull it toward or push it away from another particle, the interaction between elementary bodies such as electrons was viewed as the result of the exchange of photons. Just as two children on ice skates tossing a large ball back and forth will recoil backward, both during a toss and a catch, as the ball carries momentum between them, so do two electrons repel by tossing photons back and forth.

The attractive force between two oppositely charged bodies is a little more difficult to explain, but also more instructive. First, it must be pointed out that because of relativity and quantum mechanics particles can be created and destroyed on the subatomic scale in ways not familiar in normal experience. This is particularly true of the photon, which contains no rest energy and so is easily absorbed by and emitted from other bodies.

Imagine that the two children on skates each holds a large supply of balls and tosses them one by one in the direction away from the other. The children will recoil toward each other as the balls carry away momentum. Now the electron does not have to carry along a supply of photons; they can appear out of nothing and, after they are emitted, return to nothing. If that happens on a small enough time scale, the Uncertainty Principle allows for the small amount of energy nonconservation that occurs and which would forbid the process from taking place in classical physics.

The success of this particle-exchange picture of the fundamental electromagnetic force between bodies led to its application to the other known forces. Recall that the neutron and proton must be held together in the nucleus by a force that is neither gravity nor electromagnetism, the *strong nuclear force*. A fourth force also occurs on the nuclear scale— the *weak nuclear force*, which is responsible for the beta decay of nuclei. In 1935 the Japanese physicist Hideki Yukawa suggested that the strong nuclear force could be explained by the exchange of a particle whose mass is intermediate between the proton and electron. The proton and neutron are 1,840 times as massive as the electron. The proposed particle should be about 200 times as massive as the electron.

Next occurs one of the more curious, almost amusing, episodes in the advance of twentieth-century physics. In 1935 Seth Neddemeyer and Carl Anderson of positron fame operated their cloud chamber on Pike's Peak and observed particles apparently heavier than electrons and lighter than protons, which Yukawa interpreted as his *heavy quanta*. Similar observations had been made in England. Yukawa's interpretation was at first ignored in the West, and then found to be inconsistent with experiment. What people were seeing seemed to be another kind of electron, which was not exactly what Yukawa had proposed. Yet it had about the right mass.

The new particles, called *mesotrons*, were eventually found to occur in both positively and negatively charged varieties, like the positron and electron. Unlike the electrons, however, the mesotrons were unstable. They lived only an average of 2.2 microseconds (10^{-6} second) before decaying into an electron and some unidentified missing energy. So these mesotrons occur naturally, but with such a short lifetime that they must be constantly generated in the atmosphere by other particles from outer space, or else we would not be seeing them.

At this point, World War II intervened, slowing what had been explosive progress in science. Most physicists were recruited to design radar, the nuclear bomb, or others of the technological marvels that warfare always seems to stimulate. But in Italy three young physicists, Marcello Conversi, Ettore Pancini, and Orestes Piccioni, somehow continued their research while fighting and hiding from the Nazis and Fascists. They discovered a large discrepancy between Yukawa's theory and the mesotrons they were observing. Yukawa's mesotron, being the quantum for the strong interactions in nuclei, should have interacted more violently than they observed.

Theorists scurried around trying to explain this fact. All kinds of schemes were invented to explain how mesotrons could on the one hand interact with matter strongly enough to hold it together and, on the other hand, penetrate all the way from the top of the atmosphere where they are produced. Hans Bethe and Robert Marshak in the United States independently proposed the correct interpretation: the observed particles are not Yukawa's mediator of the strong force but a *decay product* of that particle. After the war, in 1945, a particle was observed in photographic emulsions by Cecil Frank Powell and his collaborators at Bristol University in England. This particle, now called the *pi meson*, or more commonly the *pion*, is just slightly heavier than the mesotron, which is now called the *muon*.

The pion interacts strongly with matter, while the muon does not. It is also unstable and in its charged variety decays into a muon and what we now know is a *neutrino* after an average lifetime of about 10^{-8} second. It also is found in a neutral variety, the *pi-zero*, which decays after 10^{-16} second into two photons; each photon carries off 70 million electron-volts of energy (MeV), half the pi-zero rest energy, which is in the gamma ray region of the electromagnetic spectrum.

While the pion is generally recognized as the particle Yukawa had predicted to be exchanged between protons and neutrons to hold the nucleus together, Yukawa's meson exchange theory of nuclear forces never achieved much more than qualitative success. We will return a little later to the current view of the strong nuclear force, whose under-standing awaited the peeling back of another layer of matter and the determination of the constituents of protons and neutrons—the *quarks*.

First, we must turn to the weak nuclear interaction. Recall that this force is responsible for beta decay, the emission of electrons by atomic nuclei. The simplest beta decay process is the decay of the neutron itself. Free neutrons are found to decay with an average lifetime of 15 minutes. This may seem contradictory, since we have already said that neutrons are constituents of the nucleus, and some nuclei have been around for billions of years. Again, this is a relativistic effect— the result of the equivalence of mass and energy.

The rest energy (mc^2) of a nucleus or a complete atom is less than the sum of the rest energies of its constituent particles. The difference is called the *binding energy*. To break apart a nucleus or atom, an amount of energy at least equal to the binding energy must be applied by hitting the atom or nucleus with another particle that has at least that much energy. This is fundamentally no different from boiling water where energy, in the form of heat greater than the heat of vaporization, must be applied to pull H_2O molecules out of the liquid.

It happens that a free neutron is slightly heavier than a proton and so can decay into a proton and an electron, with no extra energy needed and, in fact, some energy left over. Inside a nucleus, however, the binding energy is greater than this mass difference in most cases, exceptions being nuclei that are naturally radioactive. In stable nuclei the neutron inside does not decay because the mass, or rest energy, of the nucleus is too small to allow neutron decay to happen without violating energy conservation.

In studying beta decay, scientists found a curious result. When a nucleus decayed, the resulting electron and transmuted nucleus could be identified, yet their energies did not add up to the rest energy of the initial nucleus. Something that physicists could not detect was also being produced. There was another problem, which is best illustrated with neutron decay. The neutron and its two decay products, the proton and electron, each have a spin of 1/2, in fundamental units. It is not possible for a single spin of 1/2 to result in two spins of 1/2. Even numbers of spins of 1/2 always add up to an integer spin.

The answer was suggested by Pauli in 1930. He proposed that a very light neutral particle with spin 1/2 is being produced in beta decay. Enrico Fermi (1901–54) named this the *neutrino*, or "little neutral one." Direct evidence for the existence of the neutrino proved very difficult to obtain, since the neutrino has no charge, no measurable mass, and hardly any interaction with the rest of matter. Of course if it had no interaction at all we would not even know of its existence; in fact, it would not exist, by definition, since it would not be part of the observable universe. The only things that exist are those which produce some observable effect. The neutrino comes about as close to not existing as anything we know.

Fortunately the neutrino interacts with matter by way of the weak nuclear force, the beta decay interaction. After a decade of building and testing an experiment at a nuclear reactor, Fred Reines and Clyde Cowen were able in 1953 to report evidence for the existence of the neutrino by observing the reactions produced when neutrinos interact. Even though the interaction probability of a single neutrino is very low— one can easily pass through the earth without hitting anything—so many neutrinos are produced inside a nuclear reactor that after a long enough wait a few events can be seen.

Today neutrinos have been widely studied, with the recognition that they are among the most important particles in the universe. In fact there may be more of them in the universe than of any other particle, except possibly the photon. Although astronomers maintain that the universe is mostly hydrogen, there are a billion times as many neutrinos as hydrogen atoms in the universe. Billions pass through your body every second. They do you no harm; even the earth is almost empty space to a neutrino.

By the 1950s, the number and variety of the known elementary particles had begun to grow. Cosmic ray experiments continued with the discovery of a new class of particles that had rather strange properties and so were dubbed *strange particles*. A new quantum number called *strangeness* was assigned to these particles, more as a label than anything else, as physicists began to use more whimsical terms to combat some of the pomposity that had begun to appear among the scientific priesthood. A decade later, I wrote my Ph. D. dissertation on the interactions of one of these strange particles, the *K-meson* or *kaon*.

As the twentieth century has progressed, our view of the nature of elementary particles has changed. First, the chemical elements of the nineteenth century were replaced by electrons, nuclei, and photons. Then positrons, muons, pions, kaons, and neutrinos were added to the list. In the process, techniques changed from those of the chemistry lab to nuclear studies with naturally radioactive sources, and then to cosmic rays and nuclear reactors. Eventually, particle accelerators became the primary microscope with which to peer into matter.

This experimental progression was one of increasing energy. The reason is simple: in order to look to small distances inside matter we must use probing particles that have a wavelength smaller than the objects being examined. This can only be achieved by raising the energy of the probing particle. If these probing particles are the traditional photons, then we must raise their energy far above that of visible light, beyond even X-rays, into the gamma ray region of the spectrum. To give an idea, visible photons have an energy of about 1 eV (electron-volt). The smallest objects we can see with these are biological microbes. To "see" an atom, we need X-rays with about 1 KeV = 1000 eV; to see a nucleus we need 1 MeV = 10^6 eV; to see inside a nucleus we must go to even higher energies.

Probing particles need not always be photons. Any convenient particles may be used, and as research continued new kinds of microscopes with different probing particles of ever-increasing energy became the primary tool for the exploration of the structure of matter. The most important of the early accelerators producing these high energy particle beams was the *cyclotron*.

The cyclotron was invented in 1930 by Ernest O. Lawrence in Berkeley, California, and nuclear disintegrations produced by a beam from the accelerator were measured in that banner year of 1932. After World War II, the role that physicists had played in the development of the nuclear bomb led to a great increase in public funding for basic nuclear research, and accelerator laboratories began springing up all over the world. Technology improved yearly, and accelerators grew from the original two-inch-diameter cyclotron of Lawrence to the present-day two-mile-long Stanford Linear Accelerator (SLAC), the four-mile-circumference circular proton accelerator at Fermilab in Illinois, and the large colliding beam machines at the European Laboratory for Nuclear Research (CERN) in Geneva, Switzerland, and elsewhere. Even larger ones are being built or planned at this writing.

SLAC is capable of accelerating electrons to 0.9999999997 of the speed of light. The energy of these electrons is 20 GeV (20 billion electron-volts) and their wavelength is 10^{-16} meter. Since this length is about one tenth the size of a proton, SLAC is a giant electron microscope capable of seeing inside the proton. In 1970 physicists working at SLAC found evidence for what had already been strongly suspected: the proton and neutron are not elementary, but have substructures of their own. What they found was reminiscent of what had been observed in 1910 and had been interpreted by Rutherford as evidence for the nucleus of the atom. Just as happened with the alpha particles scattered from gold, the electrons observed at SLAC were deflected at angles that could only be explained if they were scattered from point-like particles inside the proton and neutron.

Before today's giant machines requiring vast flat territory came into operation, the hills above the University of California campus at Berkeley were the center of elementary particle physics, carrying on the tradition of Ernest Lawrence. In 1955 the Bevatron was constructed in those hills, accelerating protons to the energy of 6 GeV (which was called BeV in those days, for 1 billion electron-volts), the energy that would be needed to produce anti-protons. Everyone fully expected these to be found, and they were. The Bevatron had quickly done what it was designed to do, and then it proceeded to do what no designer could have predicted.

The Bevatron, and other proton accelerators that soon after came into operation in the United States, Europe, and the Soviet Union, began to accumulate evidence that a new layer of matter was being peeled away. As protons of ever-increasing energy were used to bombard nuclei, whole new classes of particles were produced. In addition to the strange particles already mentioned, which were first observed in cosmic rays, new extremely short-lived particles were produced at these accelerators. Some had such short lifetimes that they could barely cross the distance of a nucleus, traveling at the speed of light, before disintegrating.

By the mid-1960s, the profusion of new particles had reached epidemic proportions. Just as in the 1920s, when doubt was raised whether the ninety-two nuclei of the chemical elements could all be elementary, people began to suggest that these particles might be composed of more elementary constituents. A universe with so many elementary particles was just too untidy.

A number of physicists had been trying to find some kind of order in the particles and their properties, and sought a new Periodic Table of the elements. The most successful scheme was one proposed by Murray Gell-Mann of California Institute of Technology (Caltech). Originally called the *Eightfold Way* after the Buddhist path to enlightenment, Gell-Mann's classification for the particles now carries the less dramatic title, *SU(3)*, which stands for a particular group of mathematical operations. It had also been independently arrived at by an Israeli general, Yuval Ne'eman.

Gell-Mann did not insist that his periodic table had anything necessarily to do with a substructure of more elementary particles. He noted, however, that it did naturally follow from a picture in which there were fundamental objects of fractional electrical charge, specifically 1/3 and 2/3 of the unit electron charge. He called them *quarks*, a word he had discovered in James Joyce's *Finnegans Wake*. As we see from the epigraph to this chapter, *quark* also appears in Goethe's *Faust*, where it is translated as *trash*.

George Zweig had also arrived at the quark idea working independently at Caltech, strange as it may seem that two theoretical physicists working on the same problem at the same small university could manage to keep their ideas insulated from one another in a field where rumors can spread around the world almost instantly.

SU(3) was one of those few theories occasionally appearing on the scene that are prototypes for the way science should operate. Motivated by puzzling experimental data, it applied sophisticated but well-understood mathematical methods and made precise unequivocal predictions that could immediately be decisively tested. The *SU(3)* tables of the particles, like the Periodic Table of the elements a century earlier, had several empty boxes where no known particle fit in. The most dramatic of these openings was for a negatively charged particle that had three units of strangeness.

This prediction of new particles is most easily understood with the quark model (fig. 7.3). Initially three quarks were proposed, now labeled *u*, *d*, and *s* for *up*, *down*, and *strange*. The *u* quark has +2/3 unit charge, while the *d* and *s* quark each carry –1/3. When they combine, they always produce unit, or zero, charge. The +1 charge proton is made of two *u* quarks and one *d* quark: *uud*. The charge zero neutron is *udd*. Mesons, such as the pion and kaon, are composed of a quark and an antiquark. For example, the positive kaon consists of a *u* and an *anti-s* quark. Its strangeness is the result of the presence of the *anti-s* quark.

When we consider all the possible combinations of these three quarks and their antiquarks, we find a very strange particle—indeed, triply strange—made up of three *s* quarks. Since each *s* quark has –1/3 unit charge, this previously unobserved particle, dubbed the *omega-minus*, should have one unit of negative charge. Further, its rest energy is calculated in the theory to be 1670 MeV.

The nature of the *omega-minus* was unique; its properties were

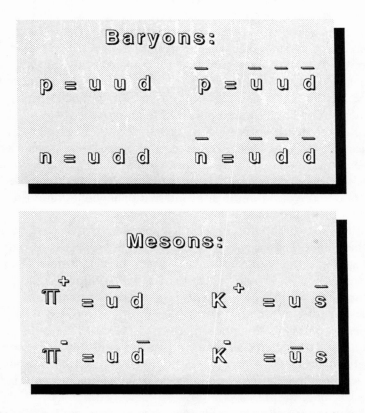

Fig. 7.3. The Quark Model. Examples are given of baryons, which are three quark combinations, and mesons, which are composed of a quark and antiquark.

unambiguously predicted by *SU(3)*. In 1964, shortly after its prediction, a particle having exactly the predicted properties was observed in a bubble chamber photograph taken in a secondary negative kaon beam at the Brookhaven National Laboratory in New York. The interpretation of the photograph was clear and unambiguous: the omega-minus was the only explanation. Yet physicists fretted that the particle was found so quickly, and they could hardly believe that they were so lucky. Thus they were somewhat relieved when the next few omega-minus particles turned out to be a bit harder to find. Physics is not supposed to be that easy!

In the years since the prediction and quick confirmation of the omega-minus, the quark theory has been very successful in explaining the properties of the large array of particles that continue to be regularly discovered at accelerator labs. Success by itself, however, is not a sufficient measure of truth. We have seen that the success of the atomic theory in chemistry did not convince everyone of its validity, because no direct evidence for the particulate nature of atoms was provided by chemical experiments. Similarly, the acceptance of the quark theory awaited some evidence of the quark's point-like nature. In the 1970s, experiments that used electrons to look inside the proton at SLAC and other electron machines, and experiments with neutrino beams at CERN and Fermilab, firmly established the quark substructure of protons. Hundreds of physicists were involved in the large international collaborations formed to conduct these difficult and costly experiments.

In the last two decades, elementary particle research has continued along these lines, gathering more details on the properties of particles, finding new particles, and testing the quark model. Rather than give a comprehensive survey of this history, I will summarize the picture of the fundamental structure of matter that exists at this writing.

The universe, as we now know it, is mostly composed of just five fundamental particles: the photon, electron, electron-neutrino, and u and d quarks. The nuclei of atoms are composed of protons and neutrons, which are made up of u and d quarks. In one view, the protons and neutrons do not maintain their identity inside a nucleus, which is instead treated as a "bag" of quarks.

What we normally think of as matter is composed of electrons and the two types of quarks. Antimatter contains positrons and anti-quarks; it exists naturally in the cosmos but in amounts a billion times smaller than normal matter.

No free quark has yet been convincingly seen, although there are experimental reports of fractional charge. In fact, the current theory of quark interactions states that quarks cannot exist as free particles outside nuclei or the other particles they inhabit. Despite this absence of direct observation, the success of the quark model is so complete that few scientists now doubt its basic validity.

Of the five elementary particles mentioned, only the photon has integer spin. Particles with integer spin are called *bosons*, and play a very special role in nature. They are not themselves components of normal

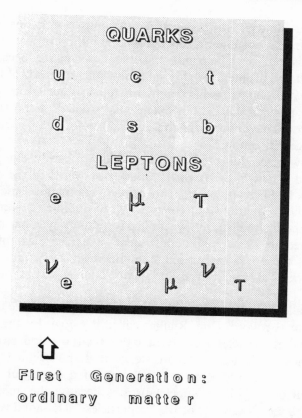

Fig. 7.4. The Standard Model. Matter is composed of three generations of quarks and leptons. Ordinary matter is composed of the quarks and electron of the first generation.

matter, but they generate the forces between the components of matter. These forces, and their unification, will be the separate story of the next chapter.

Half-integer spin particles are called *fermions*. The electron and *u* and *d* quarks that combine to make conventional matter, and the associated electron-neutrino, form the *first generation* of fermions (fig. 7.4).

A *second generation* of fermions is now well established; in fact, one member, the muon, has been around for a long time. The muon is little more than a heavier version of the electron. Produced copiously in the upper atmosphere in the decay of pions and kaons, the muon forms the major cosmic ray background on the earth. Nevertheless, the muon is not a permanent constituent of matter because of its instability. It will disintegrate after just a few microseconds of life. Short-lived *muonic atoms*, in which a muon replaces an electron in orbit around the nucleus, have been extensively studied. A possible future use, which could have enormous consequences, is low temperature nuclear fusion induced by muons.

We now place the muon in the second generation along with the *s* quark and a new *c*, or *charmed*, quark discovered in the mid-1970s. The *c* quark had been invented by theorists to cancel out some unwanted effects in their equations. I remember how dubious it seemed to us skeptical experimentalists, but we looked for it anyway. Charm originally was observed in 1974 as the constituent of a new class of particles. The first of these was discovered simultaneously by Samuel Ting and a group working at Brookhaven, and Burton Richter and a group working at SLAC.

Ting called the particle *J*, which coincidentally looks like the Chinese character for the word *Ting*. Richter called it *psi*(ψ). Further additions to this class of particles have been labeled ψ', ψ'', and so forth, but Ting has never relented in his insistence that the original particle be called *J*. He once threatened to walk out of a large public meeting with several of his fellow Nobel laureates because the announcement listed him as a codiscoverer of the "ψ particles." The nobility of science does not always extend to scientists.

The *J* or ψ particles are composed of a charmed quark, *c*, and its antiquark, $\bar{c}$. Each carries an equal and opposite value of a quantum number called *charm*, and so have net zero, or *hidden*, charm. By 1976, particles with net nonzero, or *naked*, charm were found in several experiments.

The *s* and *c* quarks combine with the first generation quarks to produce many of the unstable particles produced at accelerators. To complete the second generation of fermions, we have an associated neutrino, *the muon-neutrino*, and the corresponding antiparticles.

This is not all. There is almost certainly at least one more generation of fermions, heavier than the second, though not all the members of the third generation have been convincingly established experimentally. The electron-like particle is called the *tau lepton*. The term *lepton* is generally applied to the two nonquark members of a generation, like the electron and neutrino. The associated *tau-neutrino* has not yet been seen as of this writing. There is now good evidence for one of the quarks of the third generation, labeled *b*, and some preliminary indications for its partner, labeled *t*. The symbols variously stand for *bottom* and *top*, which matches the *up* and *down* quarks of the first generation, or *beauty* and *truth*, matching better with what most scientists like to think are the goals of their labors.

As was the case with charm, the first evidence for the *b* quark was indirect, as it appeared combined with its antiquark, the $\bar{b}$, in a particle called *chi* (χ). Thus the χ particle contained *hidden beauty*. When a meson was observed composed of a *b* and other quarks, it was referred to as *naked beauty*; some suggested the designation *naked bottom*. The search continues for the *naked top*, or *naked truth*.

In the current cold state of the universe, most of the matter we observe is composed of the first generation quarks and electrons. Except for cosmic ray muons, the second and third generations do not seem to play much of a role, only briefly appearing when particles are accelerated to high energies and go crashing into matter. This happens in the cosmos, as well as in our earth-bound particle accelerators. In fact, protons are accelerated in astronomical objects to energies as high as 10^{20} eV. These protons hitting the earth's atmosphere and producing muons and kaons gave us our first glimpse of the higher generations even before the day of the giant accelerators.

Why spend so much time discussing particles that play such an unimportant role in the universe today? The reason is that we are not so much concerned with today as with that time 10 or 20 billion years ago when the universe was young. The particles of the higher generations played a much more important role then, when temperatures and the energies at which particles collided were sufficient to produce large numbers of particles. Second and third generation quarks and leptons are apparently equivalent in every way to the familiar first generation particles, except that they are heavier. At energies that are large compared to their rest energies, as occurred in the first fraction of a second of the Big Bang, the quarks and leptons of all generations are alike and equally copious. So particles that may be rare today, except in the accelerators that produce them, were just as important in forming the universe as the current constituents of matter.

Until recently, cosmologists made the serious mistake of ignoring these particles and taking the anthropocentric view that the particles which make up earth and man have always constituted the universe. Furthermore, the study of all the generations of matter has shed light on the other great question—the nature of the fundamental forces between the particles, which gave the universe the structure that we see.

When we talk about our current knowledge of the elementary particles, a question often asked is, "What are these particles composed of?" Of course, there may be other layers of matter for us to peel back. Just as the chemical elements were regarded as the elementary particles a century ago, quarks and leptons may be seen as composite bodies by scientists a century hence. Already there are theories of the composite nature of quarks and leptons. We have no explanation for the fact that quarks and leptons exist in three, if not more, generations. These could be the various excitations of the states of more fundamental objects, called *preons*. But preon theories are little more than speculation unless they make predictions that can be experimentally tested using existing technology.

Another natural question is whether there ever will be an end to the layers of the onion. We can argue that the progression cannot be infinite; it must end at some point. If we continue to smaller and smaller distances, we will eventually reach the point where the energy of the probing particle needed to localize anything within such a small space would be greater than the total energy of the matter in the universe. Of course future scientists may find ways other than point particles to describe structure at these distances; or they may not even use our concepts of space, time, and particles as part of their description. Today's theorists are pursuing promising new avenues, such as one-dimensional *strings*, which may one day replace the early Greek concept of zero-dimensional, point-like, components of matter. Their goal is nothing less than a *theory of everything*.

8

Joining Forces

There is a gate in Japan, a gate in Neiko, which is sometimes called by the Japanese the most beautiful gate in all Japan. . . . This gate is very elaborate, with lots of gables and beautiful carving and lots of columns and dragon heads and princes carved into the pillars, and so on. But when one looks closely he sees that in the elaborate and complex design along one of the pillars, one of the small design elements is carved upside down; otherwise the thing is completely symmetrical. If one asks why this is, the story is that it was carved upside down so that the gods will not be jealous of the perfection of man.

Richard Feynman

When Newton discovered that the same Law of Universal Gravitation held for bodies falling to the earth as for planets falling around the sun, he began a quest for the unification of the forces of nature that may finally reach its goal before this century is out. We believe now that all the forces were produced as one shortly after the universe began. So if and when full unification is achieved, we will be able to write down the equations to describe that time. We cannot yet write these equations, but successful steps in that direction have been made in recent years. Further, the mechanism used to unify at least two forces, *Spontaneously Broken Symmetry,* provides the means by which accidental generation of the "laws of nature" can have come about.

By the latter part of the nineteenth century, electricity and magnetism had been recognized as two aspects of the same phenomenon. Early in the twentieth century, just two fundamental forces were recognized: gravity and electromagnetism. After publishing the General Theory of Relativity, which as we have seen is basically a geometrical theory of gravity, Einstein spent his remaining lifetime in an unsuccessful effort to unify gravity and electromagnetism by the same geometrical means, searching for a *Unified Field Theory.*

Before Einstein could bring electromagnetism and gravity together, however, the strong and weak nuclear forces were discovered, adding two more to the list of fundamental forces. To make matters worse, Einstein's General Theory of Relativity was still fundamentally Newtonian—that is, non-quantum mechanical—in framework, and physicists soon realized that any unified theory would have to incorporate quantum mechanics.

The quantum field theory that was developed in the 1940s—*quantum electrodynamics (QED)*—successfully described the electromagnetic forces between electrons in a way that the classical electrodynamics of Maxwell was fundamentally incapable of doing because it was not a quantum theory. The basic process in QED is the exchange of a photon between the interacting electrons. Related processes that are not even allowed in classical physics—such as the creation of an electron-positron pair by photon interactions with other photons, or the annihilation of an electron-positron pair into photons—can be calculated with great precision using the methods of QED.

We normally think of a vacuum as empty of matter. QED showed that a vacuum is not continuously empty, but intermittently contains electron-positron pairs, created from nothing and annihilating into nothing on very short time scales. This *pair production* and *pair annihilation* are forbidden by Newtonian physics but permitted by the Uncertainty Principle of quantum mechanics. The result is a net electric polarization of the vacuum that has real, measurable effects on the structure of atoms. The non-emptiness of the vacuum will play an important role in our understanding of how the universe could have appeared from nothing.

The great success of QED in explaining electromagnetism on the quantum scale was not immediately duplicated for the nuclear forces. Yukawa's 1935 idea that the strong nuclear force is mediated by meson exchange led to his prediction of the existence of mesons, confirmed ten years later with the observation of the pion. Despite this successful prediction and several decades of effort, the meson theory of nuclear forces simply failed to work. The reason turned out to be simple: mesons are not fundamental particles but are composed of quarks. By the 1980s a new theory of the strong nuclear force had been developed, patterned after QED and called *QCD,* for *quantum chromodynamics.* In QCD, the force that holds the nucleus together is seen to be mediated by the exchange of fundamental spin 1 bosons called, appropriately enough, *gluons.*

A key principle had to be recognized before the problem of the strong force could be elucidated: *Gauge Symmetry*. Although its full importance has only recently been realized, Gauge Symmetry was found to be present in classical Maxwellian electromagnetism by Hermann Weyl (1885-1955) in 1918. Before we discuss Gauge Symmetry, however, let us consider symmetry in general and why it plays such a profound role in the unification of forces.

A deep connection exists between the symmetries we observe in nature and the fundamental laws of physics. In fact, we now realize that the two are different expressions of the same thing. The laws of physics describe the order we observe in the universe (even if that order occurred by chance), and this order can frequently be expressed in terms of symmetry principles. Forces, as we will see, result from the breaking of symmetries.

Perhaps the most common symmetry occurs when identical objects line up at equal distances in a straight line, like telephone poles alongside a highway. In such a situation we can move the set of objects in one direction or the other in increments of the spacing without changing anything fundamental about the system. This symmetry enables us to avoid keeping a record of the position of each object; instead, we can simply specify the spacing. Thus only one number is needed where otherwise you might need thousands. For example, when laying out telephone poles along a highway, the telephone company need only specify the spacing between poles instead of the exact placement of each.

We call this *Space Translation Symmetry*. It can be extended to ideas such as the points along a line by making the incremental spacing arbitrarily small. Then the symmetry goes from being discrete to being continuous, in the sense that the points on a line form a continous set.

Extending Space Translation Symmetry from a one-dimensional line to the three dimensions of space enables us to develop these ideas further. Recall that the Copernican idea that the earth is not the center of the universe led Galileo to infer his Principle of Relativity, which in turn led to Newton's Law of Inertia. The *Generalized Copernican Principle* says that *there is no center of the universe;* every point in space is equivalent to every other point. This is not to say that the universe looks exactly the same at every point in space but that, like the telephone poles along the road, we do not have to describe the phenomena occurring at each of these points with a different set of quantities. Similarly, the laws of physics that describe phenomena will be the same at each point. If a certain physical phenomenon can happen at one point, it will happen in the same way at all other points (fig. 8.1). Newton's Law of Universal Gravitation is the same whether it is applied on the earth, the moon, or the surface of a neutron star in a galaxy far, far away.

Fig. 8.1. Space Translation Symmetry. The laws of physics are the same at different places. Here a body falls with the same trajectory in New York City as on a Pacific isle.

We would say, then, that the universe has Space Translation Symmetry. And, since no special point exists in the universe that we can call the center, there is no place absolutely at rest. Then it follows that rest and motion at constant velocity are equivalent (Galilean Relativity) and that a force is not needed to keep a body moving, but only to change its motion (Newton's Law of Inertia). So the Principle of Galilean Relativity and the Law of Inertia are intimately connected with Space Translation Symmetry. And finally, since the Law of Inertia is equivalent to the Principle of Conservation of Momentum, we have established a connection between Space Translation Symmetry and Conservation of Momentum.

We also believe that Newton's Law of Universal Gravitation is the same at the surface of a star a long time ago. A different set of laws of physics is not needed to describe the universe at different times—except, as we will see, at the earliest times. Thus the universe currently has *Time Translation Symmetry* (fig. 8.2). Just as Space Translation Symmetry and Conservation of Momentum are related, so is Time Translation Symmetry related to the Principle of Conservation of Energy. If energy, or matter, just spontaneously appeared at some time and place, an effect on the motions of bodies at that place would be observed, and we would have to describe that motion with a different set of physical laws than we used in the moment before the matter or energy appeared.

Fig. 8.2. Time Translation Symmetry. The laws of physics are the same at different times. Here a body falls with the same trajectory in the past as in the present.

The idea that the universe appears fundamentally the same at all times is also sometimes referred to as the *Cosmological Principle;* but as such it has been improperly applied to argue against the *expanding universe.* There is nothing in the concept of an expanding universe that violates Time Translation Symmetry, as long as energy is conserved. By contrast, those who have used the Cosmological Principle to argue for a steady state universe actually violate the principle by requiring matter to be continuously created in order to explain the redshift. In fact, one of the hallmarks of modern astronomy and cosmology has been that no new laws of physics beyond what we know from the laboratory must be invoked to describe the universe.

Another familiar symmetry is that of a sphere. A perfect sphere looks the same no matter how it may be rotated about any axis (fig. 8.3). This *Rotational Symmetry* is related to the *Conservation of Angular Momentum,* which is sometimes called the *Law of Rotational Inertia.* Just as a body needs an external force applied to it to change its state of motion in a straight line, a body rotating about an axis needs an external torque applied to change its rotational motion. The angular momentum L is the *quantity of rotational motion,* just as the linear momentum p is the quantity of motion in a straight line.

So there is a profound connection between symmetry principles and conservation principles. Space Translation Symmetry is connected with Conservation of Linear Momentum; Rotational Symmetry is connected with Conservation of Angular Momentum; and Time Translation Symmetry is connected with Conservation of Energy.

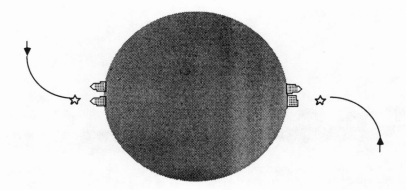

Fig. 8.3. Rotational Symmetry. The laws of physics do not depend on the orientation of the frame of reference. Here a falling body follows the same trajectory as the earth rotates in the course of half a day.

We have already remarked that conservation principles are allowed to be broken at small distances and times by the Uncertainty Principle of quantum mechanics. This implies that symmetries can be broken as well. The idea that symmetries can be broken at some tiny level was recognized in the 1950s with the discovery of *Parity Violation* in weak interactions, specifically in the beta decay of nuclei. *Parity,* in physics, is another name for *Left-Right* or *Mirror Symmetry.* While your image in a mirror does not look exactly the same as you do in real life, it does not describe an impossible configuration of human being (fig. 8.4). Similarly, the mirror image of any fundamental process involving elementary particles is also one that can occur in nature— or so it was thought until the 1950s. Experiments involving strange particles suggested to Tsung-Dao Lee and Chen Ning Yang in 1955 that parity may be violated for the weak nuclear force. Their hypothesis was tested immediately in the beta decay of cobalt by Chien-Shiung Wu and collaborators, and proved correct. The mirror images of the reaction were observed to occur at different rates (fig. 8.5).

The *SU(3)* classification of particles discovered by Gell-Mann and others in the 1960s was also recognized as a broken symmetry—one that is only approximately correct but still useful in describing what is observed in experiments. Just as other symmetries are connected to

Fig. 8.4. Mirror Symmetry. Although left and right are reversed in a mirror image, what one observes in a mirror is not generally forbidden from occurring in real life. Here a left hand appears as a right hand in the mirror image.

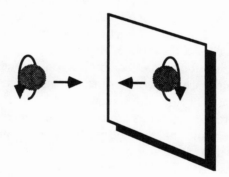

Fig. 8.5. The breaking of Mirror Symmetry. Here a spinning atomic nucleus is shown emitting an electron in the direction of a right-handed screw. In the mirror image, the electron is emitted in the direction of a left-handed screw. These processes do not occur with the same frequency, in violation of Mirror Symmetry.

conservation principles observed in nature, *SU(3)* symmetry is connected to two conservation principles. Both of these are obeyed in the strong interactions of particles; one is violated, however, in weak interactions.

The first is *Conservation of Baryon Number.* This principle says that the number of *baryons*—the heavier particles such as the proton and neutron—stays the same in any reaction. For example, in all chemical and nuclear reactions, if you were to count up the number of protons and neutrons in the nuclei of all the atoms involved, this number would not change as the reaction proceeded, no matter what changes took place among the atoms themselves, with elements transmuting to other elements.

All baryons are fermions—particles of half-integer spin—and are composed of three quarks, each having a baryon number of 1/3. Anti-quarks have a baryon number of –1/3. Mesons are integer spin bosons, composed of a quark-antiquark pair, and thus have a net zero baryon number. Consequently mesons are not baryons and so are not necessarily conserved in a reaction; there is no "law of conservation of mesons." Usually high energy reactions will produce many more mesons than exist among the initial interacting particles. Similarly, the number of photons in a reaction is not a constant; photons are continuously created in a lightbulb, for example.

Second, *SU(3)* symmetry implies *Conservation of Strangeness:* the total strangeness is the same before and after an interaction involving strange particles, when that interaction is strong. For example, when the omega-minus with three units of negative strangeness was produced in the Brookhaven bubble chamber by a kaon of one unit of negative strangeness hitting a proton with zero strangeness, the total strangeness of the system was –1. So, additional strange particles with a net +2 strangeness had to be produced along with the omega-minus. Indeed, the need for two additional strange particles to be produced was one of the signatures that allowed the experimentalists to demonstrate convincingly that they had seen the omega-minus.

An important point about strangeness conservation is that it is actually violated in the weak interactions. Strange particles decay into groups of particles whose strangeness does not always equal that of the original particle. For example, the neutral kaon with -1 strangeness decays into pions with zero strangeness. So the *SU(3)* symmetry that leads to strangeness conservation for the strong nuclear force is broken by the weak nuclear force.

The symmetries we have considered previously—space and time translation, rotation, parity—are all space-time symmetries, which we can easily relate to experience. *SU(3)* symmetry is much more abstract, operating in some *internal space* not included within the framework of our familiar three dimensions of space and one dimension of time.

Recall that a similar internal space had already been invoked with the discovery of the spins of the electron and other elementary particles. Spins represent angular momenta and so are somehow related to rotations, but these rotations cannot occur in the regular three-dimensional space of our experience, since in that space the elementary particles are points. Later we will see that the universe may in fact contain more than the four dimensions of space and time, with these additional *inner dimensions* curled up on such a small distance as not to be directly observable. *SU(3)* and the other internal symmetries are thought to operate on these internal dimensions.

Another internal symmetry, called *Charge Conjugation,* describes the fact that most physical processes take place in exactly the same way for particles and their antiparticles. Thus a more accurate name for this symmetry is *Particle-Antiparticle Conjugation.* For example, $2H_2 + O_2 \rightarrow 2H_2O$ is a chemical reaction that releases a certain amount of energy. The same amount of energy would be released in a reaction in which all the particles—nuclei and electrons—were replaced by their antiparticles. No direct evidence has yet been found for the violation of this symmetry; however, in 1964 Princeton physicists Val Fitch, James Cronin, and colleagues accidentally discovered that the symmetry combination *CP,* where *C* is charge conjugation and *P* is parity, is violated in the decay of neutral kaons.

Parity has been defined as Left-Right or Mirror Symmetry. In beta decay, the mirror image of the process does not occur with the same probability as the process itself, in violation of Left-Right Symmetry. It has been found, however, that reactions will normally occur at the same rate if the mirror image has particles replaced by antiparticles. For example, if the experiment of Wu had been done with anti-cobalt, the mirror image reaction would have been observed at the same rate as the decay of normal cobalt.

This is the case for beta decay and all other fundamental processes viewed so far, except neutral kaon decay. The violation of *CP* symmetry in this rare process, of interest only to a few specialists in what most people regard as an esoteric field with little relation to the "real world," proves to be a key to understanding how the universe came to look the way it does. Why does the universe contain a billion times as much matter as antimatter? If *CP* symmetry had been strictly true in the early stages of the universe, then there would still be equal amounts of matter and antimatter throughout the universe. Life as we know it could never have formed, as enormous nuclear explosions would take place every time an antimeteor hit the earth, or other bits of matter and antimatter collided as the normal course of events.

The observation of CP symmetry violation had another profound implication. From general principles of quantum field theory, it is possible to demonstrate that the combination CPT, where T is the *time-reversal transformation,* must be conserved in all physical processes. That is, the mirror image reaction involving antiparticles viewed in a motion picture running backward would appear the same as the original reaction involving particles viewed with the film run in the normal forward direction. In a reaction such as neutral kaon decay in which CP is violated, it follows from the CPT theorem that time-reversal is also violated. However, no direct observation of time-reversal violation has yet been made. The arguments about entropy and the arrow of time discussed earlier are not invalidated by the possibility that there may be a small breaking of time-reversal symmetry in kaon reactions. Recall that the arrow of time is just an arbitrary convention. This is still the case for virtually all practical applications, where we pay no attention to kaon decay.

So symmetries in space-time and internal space are present—and are occasionally broken—in the interactions of fundamental particles. The observation of symmetry breaking does not invalidate the usefulness of the symmetry principle as an approximation. In fact, symmetry breaking can be incorporated as part of the symmetry principle. In the case of Gell-Mann's $SU(3)$, the masses of all the particles in each of his tables are equal under perfect symmetry, but the orderly way in which the symmetry is broken enabled the prediction of the mass of the omega-minus, which was right on target.

The internal dimensions used to describe the symmetries associated with spin, baryon number, and strangeness are very abstract, and theorists reasoned their presence mainly by analogy and mathematical equivalence to space-time symmetries. Once, however, in the early universe, the dimensions of this inner space may have been as real as the familiar three dimensions of the space of our everyday experience—as real, but not as familiar. One possibility, now under intense study, holds that the inner dimensions may have been part of a universe with multiple dimensions (a current favorite is ten) of space-time, all curled up in a ball smaller than an electron. Then four of the dimensions would have spontaneously unfolded into the flat space-time we now experience, with the remaining dimensions still curled up so that they are not evident except in the way they generate conservation principles.

A more familiar conservation principle, which can also be related to an internal space symmetry, is that of *electric charge*. When scientists experiment with electricity in the laboratory, they find that total electric charge is conserved in electrical circuits and in chemical, nuclear, and particle reactions. The symmetry that leads to charge conservation is the Gauge Symmetry of Hermann Weyl. Today, Gauge Symmetry is the primary principle leading to progress in the unification of the forces.

In classical Maxwellian electromagnetism we introduce potentials that have Gauge Symmetry. Changes in the potential consistent with this symmetry lead to the same electric and magnetic fields and thus the same observable results. The simplest example is the voltage between two points, such as the plates of a capacitor. Raising the voltage level of each plate by the same amount results in the same electric field in the space between the plates and the same force on a charged particle in that space.

We now recognize that symmetries such as $SU(3)$ are, in fact, generalized Gauge Symmetries. Conservation principles in general state that quantities such as charge, momentum, baryon number, and strangeness remain constant in certain reactions. The quantum mechanical operator that represents the observable being conserved is called the *generator* of the corresponding symmetry transformation. Momentum is the generator of space translations; energy, of time translations; angular momentum, of rotations; and charge is the generator of a gauge transformation in one dimension. When the symmetry involves more than one dimension, the generator becomes a set of matrices of the required dimension, instead of a simple number like the electric charge.

Gauge Symmetry has proven to be the key to the unification of forces. Electromagnetism, which conserves charge, is gauge invariant. In the mathematical language of symmetry groups, this symmetry is called $U(1)$, where 1 refers to the one-dimensional generator of the symmetry, in this case the electric charge. The weak interaction is invariant under a generalized Gauge Symmetry, $SU(2)$, where the generators are two-dimensional Pauli matrices representing something called the *weak isospin*.

The unification of the weak and electromagnetic interactions into one force, called the *electroweak force (electromagnetoweak* would be more accurate, but obviously too cumbersome), occurred in the 1970s with the work of Sheldon Glashow, Abdus Salam, and Steven Weinberg. They put the two symmetries together into one called $U(1) \times SU(2)$. The Gauge Symmetry thus implied enabled them to write equations describing both the electromagnetic and weak nuclear forces between quarks and leptons in terms of a small number of parameters, where the basic strength of both interactions was determined by the unit electric charge.

Since the first efforts by Enrico Fermi to explain the beta decay interaction, physicists have recognized that, like electromagnetism, the weak force also might be describable by the exchange of fundamental bosons, called the *weak intermediate bosons*. The new electroweak theory required that the photon of electromagnetism must be joined by at least four new "quanta," labeled W^+, W^-, W^0 and Z^0. The superscripts of the first two represent the sign of their charges, in units of the fundamental electric charge, while that of the last two indicates neutral charge. Each, like the photon, has one unit of spin.

An important fact about the weak and electromagnetic forces discouraged early attempts at unification. These two forces do not appear at all alike in the "real world" of physics experiments. The electromagnetic force acts over great distances, while the weak force cuts off at distances of about a nuclear diameter, 10^{-13} centimeter.

The range of a force is determined by the mass of the boson that carries the force between interacting particles. The interaction takes place when one of the particles emits the boson, which then moves across to the other particle and is absorbed by it. The emission and absorption processes individually violate energy conservation. The mass of the boson is created out of nothing since the emitting particle retains its initial energy and there is no extra energy to make the boson. However, because the boson is reabsorbed by the other particle participating in the interaction, energy remains conserved overall. If the whole affair takes place in a time short enough so that the uncertainty in energy, given by the Uncertainty Principle, is comparable to the rest energy of the boson, no violation of any law of physics has occurred.

The exchanged boson cannot travel faster than the speed of light, and so can only go a limited distance in the allowed time. Thus the force has a finite range. The lower the mass of the exchanged particle, the longer the time allowed and the greater the range of the interaction. Since the photon exchanged in an electromagnetic interaction has zero rest mass, the electromagnetic force has infinite range. Actually, the force between two static electrons is observed to fall off as the reciprocal square of the distance between them. Measurements of this reciprocal square behavior of the electric force have now been done with such precision that they place an experimental limit on the mass of the photon, which is 10^{-33} of the mass of the electron. That the photon has an incredibly tiny mass, if it has any mass at all, is also confirmed by the fact that we can see light out to distances of billions of light years.

Similarly, the *graviton,* which is believed to be responsible for the gravitational force, must have zero or near zero mass to allow gravity to act over large distances.

The weak force, by comparison, acts over a range of less than a nuclear diameter. This means that the masses of the W and Z bosons responsible must be substantially greater than that of a proton. For almost three decades prior to 1980, searches had been made for the weak bosons, with no success. The reason was simple: the accelerators were not energetic enough to produce the particles. Experimentalists were largely shooting in the dark, because they did not know exactly how massive the particles would be.

This situation changed dramatically with the development in the 1970s of the concept of Electroweak Unification. The unified theory of Glashow, Weinberg, and Salam predicted exactly what the boson masses should be: the three W bosons should each have a mass eighty times that of a proton, while the neutral Z boson should have a mass ninety times as great.

Armed with this specific quantitative prediction from the theorists, experimentalists knew exactly what they had to do to find the weak bosons. The only question was finding the money and mounting the effort to do. First, they recognized that the most efficient place to look for the bosons would be in the high energy collisions of antiprotons with protons, in which beams of each are accelerated and brought together and annihilate one another. Among the debris of this annihilation would be an occasional weak boson, if the theory was correct. To produce a W boson, each beam would have to carry at least half the rest energy of the W, or 40 GeV.

By this time, protons had already been accelerated to hundreds of GeV, so protons of 40 GeV presented no severe problem. Antiprotons, however, were another story. These are not found lying around like protons, which are in the nucleus of every atom. Each antiproton must be produced by converting a billion electron-volts of energy into rest mass. So the antiprotons had to be manufactured in one accelerator and, since it takes a while to make a lot of them, they had to be stored in a ring of magnets while they were being accumulated, in such a way that they would not come in contact with any matter.

The people and equipment needed to find the weak bosons were assembled at the CERN laboratory in Geneva, Switzerland, under the leadership of Harvard University professor Carlo Rubbia. The key experimental requirement of antiproton storage was accomplished by a method called *stochastic cooling,* developed by Simon van der Meer. The rest of the experiment, while massive and costly, did not present many new technological challenges. The high-speed electronics and computers available in the 1980s could be assembled in a system capable of recording and analyzing data at extremely high rates. The result was complete success.

In 1983 a paper appeared in *Physics Letters* written by Rubbia and van der Meer, along with 134 co-authors, announcing the observation of the W boson at a mass of 80.8 GeV, exactly as predicted. This was followed shortly by the announcement by the same group of the observation of the Z boson at 92.9 GeV. These results were soon confirmed by another independent group, also working at CERN. The electromagnetic and weak interactions had finally been verified as manifestations of the same fundamental force. The Nobel prize committee had been so sure of this that they had already given the award to Glashow, Weinberg, and Salam in 1979, four years before. Rubbia and van der Meer received theirs in 1983, the same year as their publication, in one of the quickest awards on record.

Electroweak Unification is truly remarkable in that it says five bosons are responsible for both the extremely long-range electromagnetic interactions and the extremely short-range weak interactions, and that these bosons are really just different states of a single fundamental boson. For a while many physicists found it hard to believe that the massless photon is equivalent to W and Z bosons, which are heavier than silver atoms. Many still are incredulous, but there is no denying the success of Electroweak Unification. Currently not a single piece of experimental data is inconsistent with the quantitative and qualitative predictions of the theory.

Even physicists sometimes have problems of perspective. The eyes we were born with are able to detect photons, which are everywhere and require little energy to produce. By contrast, it took the massive effort of hundreds of scientists, engineers, and technicians and large amounts of money to achieve for a tiny instant the energy required to manufacture weak bosons in the lab, and an almost equal effort to detect them. Such energies are not common on earth; the temperature that would be required to produce this much energy thermally is 5×10^{14} degrees Celsius! In the world of our experience, there is little in common between photons and weak bosons.

But let us imagine a world where such temperatures are commonplace. Then the difference in mass between a photon and the weak bosons would be negligible, and the interactions produced by their exchange would be indistinguishable and long-range. In such a world, weak bosons and photons would be produced in particle collisions with comparable likelihood. At sufficiently high energies, the electromagnetic and weak forces are unified and the $U(1) \times SU(2)$ symmetry that describes this unity is perfect.

There once was a time when very high temperatures occurred and higher levels of symmetry were exposed: in the fraction of a second after the birth of the universe. In the cooler world in which we live the symmetry is broken, the way icebergs break the smooth planar symmetry of an arctic sea, and differences between the bosons appear (fig. 8.6). Only when we achieve the high energies required in a particle collision at an accelerator can we briefly glimpse the underlying symmetry today.

Recall that the meson-exchange model for the strong interaction, first proposed in 1935 by Yukawa, did not prove viable. However, suc-

Fig. 8.6. Structural order as broken symmetry. Here the phase transition of the freezing of water into ice breaks the planar symmetry of an arctic sea, resulting in a more structured but less symmetric iceberg.

cesses with the Gauge Symmetry for the electroweak interaction have been almost matched by a successful gauge symmetry description of the strong force. First, the strong force only acts between quarks; leptons are limited to the electroweak force, which also acts on quarks. Of course gravity also acts on all particles, but let us ignore it for now.

We have discussed how the generator of the $U(1)$ symmetry of the electromagnetic force is the one-dimensional charge and how the generator of the $SU(2)$ symmetry of the weak force is the two-dimensional weak isospin. It was found that the strong interaction obeyed an $SU(3)$ symmetry, in which the generator is a three-dimensional mathematical entity called the *color*.

Unlike strangeness or charm, color is not simply a whimsical name given to a property of particles. Color, as used here, is analogous to our familiar primary colors. The three dimensions of the color $SU(3)$ symmetry are like the primary colors red, blue, and green. They combine to form other colors, but equal amounts of the three result in white, or the absence of color. A quark has one of the three primary colors. When three quarks combine to form a fermion, such as a proton or omega-minus, each has a different color and the net adds up to white. An antiquark has one of the colors complementary to red, blue, or green. An antifermion, such as the antiproton, would be white. Mesons, composed of quarks and antiquarks, also are white, since a color and its complement add to white. In other words, all the particles we actually observe as free particles are colorless, or white.

Particles such as quarks that have color do not appear to occur freely but are confined to existing inside familiar fermions and mesons, at least at the energies we normally study. Unlike electromagnetism, the force between quarks grows as they are pulled farther apart. When this happens, such a strong field is generated that it breaks down, like a lightning bolt between two highly charged clouds, and new quark-antiquark pairs are generated along the line between the original quarks, analogous to the way ions form along the path followed by a lightning bolt. These quarks and antiquarks then combine with the original quarks in colorless combinations so that we never see the colorful quark all by itself. Instead a swarm of mesons is produced in the interaction. Just as an electrical discharge such as a lightning bolt acts to neutralize electric charge, the color force acts to neutralize color.

So color plays the role of charge in the strong interaction. This prompted Richard Feynman to dub the theory quantum chromody-namics (QCD). As mentioned, the mediator of the color force in QCD is a unit spin gauge boson called the gluon. Eight of them are required, since they must carry color between quarks in all the ways that a quark and antiquark of any color can combine. The three possible primary colors of the quark can combine with the three possible complementary colors of the antiquark to give a nonwhite gluon with six possible color combinations. In addition, there are two independent ways to produce white, for a total of eight differently colored gluons. Most physicists believe that QCD gives a correct description of the strong nuclear force, although experimental tests are not yet conclusive.

The attempt to unify the strong and electroweak interactions into a single force is the natural next step in the sequence: the *Grand Unification Theory,* or *GUT.* From Electroweak Unification, we know how to seek the next synthesis. We want to look for an internal Gauge Symmetry that exists at higher energies and breaks down at lower energies into the $U(1) \times SU(2)$ symmetry of the electroweak force.

A common characteristic of the internal symmetries so far discussed is the way they group different particles together in a single unit. The first attempt at this kind of grouping was made many years ago for the proton and neutron. These *nucleons* have almost identical masses, suggesting strongly that each is a different state of the same particle. The symmetry that combined the two was called *isospin,* by analogy with spin angular momentum, described by the group *SU(2).* In fact the weak isospin of the weak interaction was introduced by analogy with nuclear isospin. This two-dimensional internal symmetry was later expanded to a third dimension with *SU(3),* which grouped particles with quite different masses, such as the proton and the *lambda hyperon,* and then again with Electroweak Unification, which grouped the massless photon with the weak bosons, the heaviest of the known elementary particles.

In GUT the eight gluons combine with the photon and four weak bosons, becoming different states of one or more X bosons. Similarly, the distinction between quarks and leptons breaks down. Thus we obtain a system with only two sets of elementary particles: integer spin gauge bosons and half-integer spin *leptoquark* fermions. The leptoquarks build up matter, while the gauge bosons produce the forces. This all happens at some *unification energy,* estimated to be at about 10^{14} GeV or higher.

Since 10^{14} GeV is far above anything we can hope to produce in the lab, we might have had no way to test this concept. However, one fundamental implication could be tested: GUT implies that protons are unstable, ultimately decaying into lighter particles such as electrons and mesons.

To understand how any effect of a symmetry that occurs only at energies above 10^{14} GeV can possibly be observed, recall that a correspondence exists between high energy and small distance. For particles to be localized at a given distance, their de Broglie wavelengths must be small compared to that distance. At small distances this wavelength must be smaller still, and the momentum and energy of the particles must be large; the smaller the distance, the larger the energy. Now, just as we use high energies to probe small distances, we can use small distances to probe high energies.

A particle with an energy of 10^{14} GeV traveling near the speed of light will have a wavelength of 10^{-29} meter. At distances smaller than this inside a proton, the masses of the constituent quarks are negligible. In this incredibly tiny region of space, a higher level of symmetry exists, with quarks and leptons indistinguishable. Thus quarks inside protons have a small probability of becoming leptons, resulting in protons disintegrating into mesons and leptons.

Of course, this must happen with a very low probability, or else the world would look very different from the way it does. The mean lifetime of a single proton is known to be much larger than the age of the universe, just by the fact that we are still here, with all our protons. Recent experiments, which do not wait for a single proton to decay but patiently keep watch on a very large number of them, have failed to produce any convincing examples of proton decay, though occasional "candidates" are reported at conferences. The current limit on the lifetime of the proton is 10^{31} years, accurate enough to already rule out one of the GUT theories that had been proposed—*minimal SU(5)*, the simplest of the large number of ways to achieve Grand Unification. The discovery of proton decay with the lifetime predicted by *SU(5)* would have been a truly remarkable event. Three of the four known forces would have been unified in just one decade. But this was not to be.

Grand Unification remains, at this writing, a promising but unfulfilled hope. Many physicists now believe that the answer to further unification does not lie in the strong and electroweak forces alone but in unifying these along with gravity. Quantum gravity is also a gauge theory, but the gauge bosons have two units of spin rather than one. This complication discouraged research for a while, but a great effort is now in progress to bring gravity into the fold. Most approaches center around the idea of *supersymmetry*.

To bring the spin-2 graviton into the same fold with the spin-1 photon and weak bosons requires that spin be incorporated into the symmetry structure. GUT would unify the spin-1/2 leptoquarks into one class of fermions, and the gluon, photon, *W,* and *Z* into another class of bosons, but would keep these two classes of particles separate.

In supersymmetry, the fermions and bosons are combined into one *superparticle.* One result is the prediction of a whole new class of elementary particles, which are the *supersymmetric partners* of the known particles. For every known boson, there is a related supersymmetric fermion, and vice versa. The spin-1/2 electron has a spin-0 partner, called the *selectron.* The spin-1 photon has a fermion partner called the *photino.* There are *sneutrinos* and *gravitinos* and, in one of my favorite designations, there are supersymmetric partners of the quarks called *squarks.* Unfortunately, despite exhaustive searches at accelerators, no evidence has been found for supersymmetric particles. Perhaps the accelerator energies are not yet high enough.

Whatever the ultimate unification scheme, if any is ever found, the question still arises as to the nature of the inner dimensions that correspond to spin and the other quantum mechanical observables, such as baryon number and strangeness. As early as the 1920s, Theodore Kaluza and Oskar Klein suggested that a fifth dimension be added to the General Theory of Relativity to account for electromagnetism in the Unified Field Theory that Einstein and others were seeking. As discussed earlier, the general relativistic approach to unification was doomed to failure, since any unified theory would have to be a quantum theory, which general relativity is not, and encompass the nuclear forces not yet known at that time. Recently the Kaluza-Klein idea of higher dimensions of space-time has been brought back into the picture as a possible way to explain the *internal symmetries.* Like supersymmetry, this idea is still highly speculative and may turn out not to work, but it has a certain appeal.

The traditional view is that three dimensions of space and one of time exist, consistent with our common sense. Suppose, however, that there are really more than four dimensions. The additional dimensions have such a high degree of curvature that the axes fold back on themselves in a distance small compared to a nuclear diameter.

Einstein showed that space can be thought of as curved, so that if you start out in one direction you will eventually come back to the same spot, if the curvature is positive. This occurs on the two-dimensional surface of the earth, so why not in higher dimensions, and why not on a subnuclear scale? It appears that at least ten dimensions are needed to provide all the degrees of freedom of the known particles and forces.

Another line of attack that is currently being intensely studied is closely related to both supersymmetry and higher dimensions—*superstrings*. In this approach to unification, the fundamental constituents of the universe are not the point-like particles envisaged by the Greeks, but tiny curled-up strings that appear as point particles at the energies of existing probes. This is currently the hottest theoretical fashion in particle physics. We cannot predict at this time whether superstring theory will eventually disappear, like most fashions, or become, as its proponents suggest, the *theory of everything (TOE)*. Let us wait and see.

Whatever the ultimate fate of the latest lines of study, a revolutionary new way to describe the order we observe in the universe has been achieved and is likely to survive. First, what we call the laws of nature can be expressed in terms of symmetry principles in a multidimensional framework that includes space, time, and some inner dimensions. Second, the higher the energies of the particles involved or, equivalently, the smaller the distances, the higher the level of this symmetry. At higher energies and smaller distances, the world becomes more symmetric, which means it is simpler to describe: fewer fundamental objects (particles, strings, or whatever); fewer parameters needed for the description; and fewer forces.

As we move from small distances to larger, or high energies (temperatures) to lower, the symmetries begin to break down and the world becomes more complicated to describe. This is precisely analogous to what is observed in common experience as matter cools from high temperatures to low. When a gas liquefies or a liquid solidifies, it passes from higher to lower symmetry in what is called a *phase transition*. A gas uniformly fills a container, with its molecules equally likely to be anywhere. When it liquefies the molecules will gather in a spherical shape, unless the experiment is performed in a gravitational field as on the surface of the earth, in which case it collects at the bottom of the container. When the temperature is lowered further, the original gas will have become a solid that may have a complex and beautiful shape, such as a snowflake, but still possesses less overall symmetry.

This is the stage that fundamental physics has reached as we approach the end of the 1980s. If we cannot yet describe the world by means of one theory of everything, we at least now know enough to begin to grasp how the universe that we live in, with its particular set of rules and structures, came to be.

9

The Vale of Broken Symmetry

Think binary. When matter meets antimatter, both vanish, into pure energy. But both existed; I mean, there was a condition we'll call "existence." Think of one and minus one. Together they add up to zero, nothing, nada, niente, right? Picture them together, then picture them separating—peeling apart. . . . Now you have something, you have two somethings, where once you had nothing.

John Updike, *Roger's Version*

What is beauty? While our concept of beauty may be largely subjective, certain aspects seem to be universally accepted. Few people would not regard a flower or a sunset as beautiful. Rough, scaly skin is ugly, while smooth, clear skin is beautiful. A random jumble of street sounds is noisy, while the rhythmic whistle of a bird is beautiful. One might conclude that beauty is order or symmetry, and ugliness is disorder or randomness. But is a straight line beautiful? A flat plane? A perfectly smooth sphere? Do we find the constant repetition of the same musical note beautiful? Quite the contrary. It is, in fact, the little imperfection, the deviation from perfect symmetry, that we usually recognize as beauty.

Beauty is *broken symmetry,* a kind of superposition of randomness and order. A beautiful tree has a certain symmetry, but the almost random placement of its branches and leaves gives it the interesting complex appearance we find so appealing. An electronic oscillator can provide a far purer note than any musical instrument, but the imperfect resonances of the instrument provide a richness of sound that computer-controlled oscillators can only copy and, so far, not invent. A great musical composition has its cadence and its motifs, but unexpected changes that take place at various points in the piece make it far more interesting than if they had occurred in predictable fashion.

Similarly, the snowflake formed by the freezing of a rain droplet is more beautiful than the droplet, though the latter has the higher symmetry. And the droplet is more beautiful than the water vapor out of which it formed. So beauty is not perfect symmetry; nor is it perfect order. And if it is necessary for physicists to presuppose a certain amount of chaotic behavior in describing the universe, because the data require it, this does not detract from the beauty of the description—but indeed enhances it. The completely deterministic, mechanical picture of the universe is not correct, and neither is it beautiful.

We have seen the deep role that symmetry principles play in our current description of the fundamental structure of the universe. The impression may have been left that symmetry is synonymous with order, but this is not always the case. A high level of symmetry can be either very orderly or very disorderly, or someplace in between. Evenly spaced telephone poles are orderly and symmetric, but the molecules in a gas are disorderly yet also highly symmetric.

What we mean by order in the description of physical phenomena is *predictability*. What we mean by symmetry is *simplicity* of description. The motion of a planet is both predictable and symmetric; its orbit is an ellipse lying in a plane and can be defined by just two numbers, the major and minor axes. Knowing these, and the laws of gravity and motion, we can predict where the planet will be at any given time in the future.

A container of gas is also very symmetrical, in the sense that it can be simply described by a few numbers: the volume, pressure, and temperature are adequate in most cases. But the motion of the individual molecules of the gas is highly unpredictable. While we can calculate the average number of particles that might be found at a particular place in the volume at some time, we find it impossible to predict where an individual molecule will be because of the great number of particles involved.

The universe today is on the whole both less orderly and less symmetric than it was at the beginning, although there are tiny pockets of highly ordered structures such as our earth and its life forms. While our eyes and telescopes present us with a fairly orderly picture of planets, stars, and galaxies, the quarks and electrons that constitute them are no more than one part in a billion of the particles in the universe. The overwhelming majority of these particles exist in a highly disordered state.

There are about 10^{79} quarks and electrons in the atoms that compose the visible matter of the universe. By contrast, the 2.7° microwave radiation first observed by Penzias and Wilson in 1964 is composed of about 10^{88} photons. That is, there are a billion photons for each atom of ordinary matter. These photons appear to be so uniformly distributed that the temperature measured in all directions from the earth is the same to within one part in 10,000. They constitute a volume of gas that is highly symmetric, but also highly disordered.

Recall that *entropy* is the quantitative measure of disorder used in physics. In the case of a gas of photons, or other relativistic particles, the entropy can be simply equated to the total number of particles. This is a sensible measure of entropy since the greater the number of particles present in a system, the more disorderly that system becomes when those particles move about randomly. An analogy might be a dormitory room full of college students: the more students in the room, the closer it comes to utter chaos.

Since there are a billion times as many photons as particles of ordinary matter in the universe, the entropy of the galaxies and their suns and planets is tiny compared with the entropy of the photon background. So the entropy of the universe is currently at least 10^{88}, and may be even higher if there is a component of undetected *dark matter,* as cosmologists now believe.

The number of photons in the background radiation was not 10^{88} at earlier times, since the total number of photons in the universe is not a constant. Photons have no rest energy and so are easily created and destroyed. For example, the thousands of photons that stream from a lightbulb after it is turned on did not exist in the bulb prior to applying power. If the absolute temperature of a photon gas is T, then the average energy of a single photon will be $3kT$, where k is Boltzmann's constant. If this gas expands without energy being added or subtracted, it will cool. The constant total energy of the gas will then increasingly be divided up among more particles, each receiving less of a share of energy than before. So, as the universe expands and cools, the total number of photons in the background increases. Currently the temperature is 2.7° Kelvin, and the number of photons is 10^{88}. At an earlier time when the temperature was less, the number was lower. For example, when the temperature was 27,000°, there were 10^{84} photons in the background.

Despite the high entropy of the photon background, it is still not anywhere near the maximum possible entropy the universe can have at any time. We can understand this as follows: Nothing can have higher entropy than a black hole, since the states of matter within a black hole cannot be observed and so are completely uncertain. Since entropy is disorder, or lack of information, the situation of minimum information is that of maximum entropy.

Thus the maximum possible entropy of the universe is the entropy of a black hole of the same size. The entropy of a black hole can be calculated from a formula derived by Stephen Hawking. This says that the entropy is essentially equal to the number of photons that would be produced if that black hole were to disintegrate into photons.

Hawking showed that a black hole is fundamentally unstable, ultimately disintegrating with a mean lifetime that increases with its mass; stellar-sized black holes have lifetimes larger than the age of the universe, but small black holes can have a very short lifetime.

Recall that prior to the Planck Time, 10^{-43} second, it was impossible to distinguish between particles, or even define the concept of a particle, since the de Broglie wavelength of a particle would have been larger than the size of the universe at that time, 10^{-33} centimeter. Thus in some sense, the whole universe was a single particle at the Planck Time, with an entropy equal to unity, the maximum value possible. In fact, this single-particle universe would have been a black hole with a radius of 10^{-33} centimeter, and the Big Bang may be viewed as the Hawking disintegration of this black hole.

This conclusion has enormous implications. Although the entropy of the universe at the Planck Time was at least 88 orders of magnitude *lower* than it is now—meaning that the universe now is 88 orders of magnitude less orderly than it was then—the universe at the Planck Time nevertheless must have been in complete chaos! This follows because the entropy was as great as it could possibly be for an object of its size, and so the disorder of the universe *at that time* could not have been any higher. There was no room in the entropy of the universe at 10^{-43} second for any organization. *There could not possibly have been any grand design at that time.* Or, put another way, if we are to conclude that there was an organized plan to the universe prior to 10^{-43} second after the Big Bang, then we must violate some of the principles of physics, which in all other cases are consistent with what we know about the universe. Far from explaining things, the idea of a creation forces us to seek new explanations for the world around us.

In the spirit of Occam's Razor, since nothing has yet been observed about the universe that is inconsistent with the principles of physics,

until some crucial inconsistency is found, we must tentatively conclude that the universe started out in complete chaos, with its entropy already maximum.

How then is it possible that order formed, consistent with the Second Law of Thermodynamics? If order is to form, does the total entropy of the universe not have to increase, or at least remain constant, while the local entropy of the ordered system decreases?

This seeming paradox is resolved by the fact that the universe is expanding. At the Planck Time, the entropy of the universe was maximum. As the universe expanded, its maximum allowable entropy, given by the entropy of a black hole of the same size, grew far faster than that of its main components. In other words, the universe existed in a condition of maximum entropy as long as it was a black hole. Once it exploded, however, it became an expanding gas with less than its maximum allowable entropy (fig. 9.1).

The universe can become more disorderly as a whole while it develops order in its various parts. The maximum allowable entropy of the ever-expanding universe continually increases at a rate far faster than the background photons, with plenty of entropy left over for orderly structures to develop. The local entropy of these structures decreases while the total entropy of the universe still increases or remains constant, satisfying the Second Law of Thermodynamics. In fact, the Second Law of Thermodynamics is nothing more than the statement that events happen on the average in the direction of their most likely occurrences; so the order that resulted after the Big Bang is not some highly improbable miracle but just the way the dice fell. The decrease in entropy needed to produce the earth and keep it going, or even 10 planets for every star in the universe, can easily be compensated by an increase in the total entropy of the photon background, which was at least 20 orders of magnitude below its maximum allowable amount at the time the galaxies were formed.

Earlier, I argued that order can occur by chance; that no agent is necessary to produce order, although such agents do exist. When we look at the universe as a whole, however, it appears that *no agent was possible* since the universe began in total chaos. The presence of the words on this page, however, shows that order now exists, so let us proceed to see what may have happened when a black hole of 10^{-33} centimeter exploded to produce the universe.

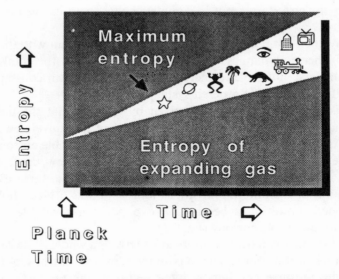

Fig. 9.1. Formation of order in the universe. At the Planck Time (10^{-43} sec) the universe had maximum allowable entropy. As the universe expanded, its entropy increased more slowly than the maximum allowable, leaving an ever-increasing band for the formation of order.

What exactly is that universe? Carl Sagan, in his "Cosmos" television series, told us that the universe is composed of "billions and billions" of galaxies, each with "billions and billions" of stars. But that describes only the universe of visible light, not the part that really counts. We have seen that the primary component of the currently observable universe is the photon background. Besides these microwave photons, a comparable background of neutrinos left over from the Big Bang almost certainly exists, although difficult or impossible to observe with current technology. Some have speculated that these neutrinos might have a very slight mass, about 6 percent of the mass of the electron, so far unconfirmed experimentally. Even if the mass of an individual neutrino is only about 1 percent of the mass of an electron, the number of neutrinos is so great their total mass would be sufficient to provide 90 percent of the mass of the universe. This scenario is dubbed the *neutrino-dominated universe.*

Considerable evidence for a large hidden mass of matter in the universe has been accumulating in the last few years. This dark matter is not seen with our telescopes because it does not radiate light in any of the spectral regions explored in conventional astronomy. Whether the dark matter is composed of neutrinos or some other particles is unknown and the subject of considerable study at this time. Current calculations do not favor neutrinos, but rather some form of *weakly interacting massive particles,* dubbed *wimps,* not yet observed. Thus the most important, in fact dominant, particle in the universe may not yet even have been discovered!

Whatever the dark matter consists of, the visible stars and galaxies are really a small part of the universe. They are a sort of afterthought— a little debris left over when the universe formed. Starting out in a highly disordered but symmetrical state, the universe expanded and cooled, and various particles and fields were generated. At some hot early stage there would have been equal numbers of particles and antiparticles as these annihilated one another and, in turn, were produced in pairs by the interactions of photons and other particles. When the universe had cooled to the point where the symmetry between particles and antiparticles was ever so slightly broken—by one part in a billion— an excess of particles over antiparticles resulted. When that happened, a sufficient number of antiparticles no longer existed to annihilate all the particles, and 10^{79} or so quarks and electrons were left over to coagulate into the galaxies.

So when Fitch and Cronin accidentally discovered in 1964 that particle-antiparticle symmetry is broken in the decay of neutral kaons, they unknowingly found the key to understanding one of the long-standing puzzles of cosmology—why there is so much more matter than antimatter in the observable universe. Previously, the asymmetry between matter and antimatter had had to be taken as an *initial condition,* generated in the ratio of 10^9 to one when the universe was "created," with no satisfactory physical explanation.

The resolution of what is called the *baryon asymmetry problem* had an important effect on our search to understand how the universe came to be. It demonstrated a close connection between our knowledge of the very small, as pursued in elementary particle physics, and the very large, as pursued in cosmology. The early universe was found to be a high energy physics laboratory. Particle physicists and cosmologists began to talk with one another. Previously, the latter had confined their use of microscopic physics to atoms and nuclei, and the former had mostly worked with data from earthbound accelerators.

You may wonder how scientists can presume to apply the knowledge gained in laboratories to a time 10 billion years in the past. Is this not simple arrogance? We of course are making an assumption that our scientific methods apply to the study of the early universe. The alternative is to relegate the most fundamental issue in nature to irrationality, ignorance, and superstition. The detailed knowledge of forces and particles that we now possess extends to energies that existed only 10^{-16} second after the Big Bang. This already takes us back to the time before matter as we normally know it was generated. Further, we have no reason to doubt that our general principles of relativity and quantum mechanics apply down to the Planck Time. So our speculations rest on a much firmer basis than might offhandedly be assumed.

In the early universe, temperatures were so high that atoms and nuclei were stripped down to their fundamental parts: quarks, leptons, or any constituents of these that may exist. Moreover, the broken symmetries of elementary particle physics can account for the asymmetries we observe in the universe today. These asymmetries do not have to be assumed as initial conditions; they can be accounted for as spontaneous processes that occur as the universe expands and cools.

While the 2.7° microwave background is highly uniform and isotropic, the same is obviously not true for the matter we see around us. In space, matter clumps into stars, planets, and comets. On earth, it clumps into clouds, mountains, trees, and people. The stars gather in galaxies, which also form groups. Recent observations on the largest scales indicate that visible mtter is far from uniform, with gigantic gaps of empty space millions of light years in extent. A very active field of study these days uses supercomputers to simulate the way density fluctuations in the early universe could have grown to the asymmetric patterns we see.

The primordial density fluctuations that triggered the clumping process seem to require the presence of weakly interacting particles in great numbers. The neutrinos left over from the Big Bang were an early candidate for the source of the fluctuations. However, more massive particles now appear to be needed, perhaps the same particles, the wimps, that may constitute the dark matter.

The exact details of galaxy formation are really not critical to our main issue of universe formation. The galaxies happened much later. Our interests here focus on an earlier time, on the heart of the Big Bang rather than on the unburned remnant left over from the explosion. Let us go back to the time before this residue appeared, when the temperature was so high that any particles which may have been present moved at speeds very close to the speed of light. At this primordial time, the universe existed in a highly symmetric state within a tiny volume a fraction of the size of an atomic nucleus. As this volume expanded and cooled, it underwent phase transitions in which the symmetries were broken and the particles, forces, and other physical laws we now observe became frozen into place.

The most familiar phase transitions in matter are the changes that occur between the gas, liquid, and solid states, although many others are studied in what is now called *condensed matter physics*. Phase transitions are characterized by an abrupt change in entropy. For example, when a raindrop freezes to a snowflake, its entropy decreases with the production of a more orderly structure.

A better analogy with the early universe is the example of a ferromagnet (fig. 9.2). All materials contain atoms, which themselves are tiny bar magnets, like compass needles, by virtue of their orbiting electrons. At high temperatures these atoms have large random motions, so that the compass needles point in all directions with no net magnetization in the material. As the material cools, this motion lessens and various subregions, or *domains*, containing large numbers of atoms tend to lock into place. For most materials the orientations remain random, but for iron and other magnetic materials, at some critical temperature called the *Curie point* a phase transition to a more orderly state is energetically favored (fig. 9.3). Below the Curie point, the domains become shaped so that the compass needles tend to point in the same direction, their fields then reinforcing one another to produce the large net field of the magnet.

The production of magnetization below the Curie point constitutes a breaking of symmetry, as the magnet's orientation selects a particular preferred direction in space where previously none existed. But note that nothing chooses this special axis ahead of time. The resulting magnetic axis can be oriented in any direction; the particular one chosen is purely accidental. Thus while the magnet itself does not exhibit the full underlying symmetry of the process that produced it, that symmetry

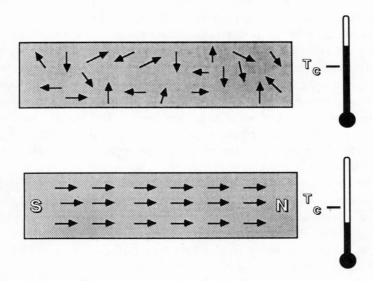

Fig. 9.2. Transition to order in a ferromagnet. Above a critical temperature, T_c, the atoms inside a bar of iron are randomly oriented and there is no net magnetic field. Below the critical temperature, the atoms line up and their individual magnetic fields add to give a net magnetization to the iron bar.

is nonetheless inherent, as can be seen from the fact that all directions were initially equally probable. In examples such as this, we say that the symmetry is hidden. The magnetic force is symmetric, despite the observed asymmetry of magnets.

The specific particles and forces that compose our universe result from broken symmetries that analogously select out particular directions in a multi-dimensional space, as our universe expands and cools and these directions freeze in. Each particle state, or unique force, is represented by a vector, like the compass needle, in this space. For example, an *up* quark points up and a *down* quark points down in the weak isospin space of the quarks.

Like the orientation of the magnet, the directions that were ultimately chosen in each phase transition in the early universe were accidental. Thus the specific set of particles and forces we now observe, and the rules they obey, are the result of chance. As we will shortly see, our

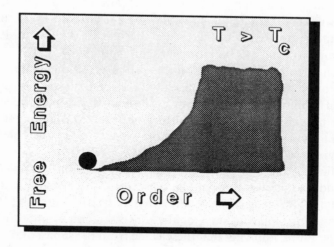

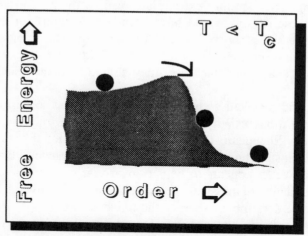

Fig. 9.3. Phase transition to a more orderly state. In the top figure, the temperature, T, is above the critical value T_c, and the system, symbolized by a ball, resides at the bottom of a valley with minimum free energy and order. In the bottom figure, the temperature has dropped below critical and the system rolls down to a new low energy state of higher order.

universe is very possibly just one domain in a vast number, each with its unique structure and symmetry.

The idea that the universe may have formed spontaneously from nothing is an old one, appearing in many creation myths and suggested by many thinkers throughout history. It first resurfaced in the modern scientific literature in a short article published in *Nature* in 1973 by E. P. Tryon (Tryon, 1973). However, this speculative paper was not given much attention. At about that time, though, serious attempts were beginning to be made to mathematically describe the events of the early universe. Theoretical physicists and cosmologists gradually recognized that problems existed with what was then the accepted Big Bang scenario, in which Einstein's cosmological equations were used to extrapolate back to *time zero* at constant entropy.

The first of these is called the *Horizon Problem:* The microwave background radiation is smooth and isotropic over regions of space that could not have been causally connected in the early universe. Regions of space now on opposite sides of the universe contain photons of the same temperature, implying that these came from the same source during the Big Bang. If we simply extrapolate back with the current rate of expansion, these regions would have been too far apart for any signal traveling at the highest possible speed, the speed of light, to pass between them in the few seconds that the universe had been in existence up to that time. That being the case, they should not have such similar properties today.

The second problem with the standard Big Bang scenario is called the *Flatness Problem:* Why is the geometry of the universe today so close to being Euclidean, that is, the three-dimensional equivalent of a flat plane? Einstein had shown that space would in general be expected to be curved. Yet it appears to be very nearly, if not completely, flat— except in the vicinity of massive objects such as stars. The overall curvature of the universe determines whether the universe is *open* or *closed*. This curvature depends on the density of matter averaged over the whole universe. If the average density is below a certain critical density, about 10^{-29} grams per cubic centimeter, the universe will continue to expand forever. If the average density is above this value, the universe will eventually stop expanding and ultimately contract back to a point.

Astronomical measurements of the average density of the universe have tended to favor the open universe, with estimates consistently lower than the critical density by about a factor of 100. Other estimates, based on calculations of the synthesis of light nuclei in the Big Bang, indicate that the density of normal matter—that is, matter composed of familiar nucleons and electrons—is about 10 percent of the critical value. The surprising thing about these results is how close the universe comes to having exactly the critical density required for a delicate balance between being open or closed or, in terms of geometry, exactly flat. The curvature of the universe will evolve with time, unless the universe is exactly flat. This fact can be seen easily in the analogy of an expanding balloon, where the curvature of the balloon's surface decreases as the balloon expands. On the other hand, a flat piece of rubber remains flat as it is stretched. For the universe today to be even a factor of 10 or 100 from being flat required a fine-tuning of the parameters of the early universe of 55 or so orders of magnitude.

Again, this is an initial condition of the universe that one would like to explain. The most natural explanation is that the universe is indeed flat. The dark matter mentioned earlier could very well provide the missing mass. If so, then 90-99 percent of the universe has not yet been discovered! Furthermore, this dark matter is not some distant component, far out of sight of our telescopes, but everywhere in the universe including right here on earth.

The third problem with the Big Bang scenario is called the *Monopole and Domain Wall Problem.* This was more of a problem for the elementary particle physicists trying to apply their ideas to the Big Bang than to cosmologists. The physicists found that the universe implied by their equations was dominated by geometrical structures called *topological defects.* These defects are similar to those which occur within crystals. Point-like defects are called *magnetic monopoles.* One-dimensional defects are called *strings.* Two-dimensional planar defects are called *domain walls.*

Monopoles can be viewed as a situation in which the conventional dipolar magnetic domains within a ferromagnet line up like the points of an asterisk (*), with their tails at the center and their arrows all pointing radially outward. The magnetic field that would result in this case would look just as if it had come from an isolated magnetic point charge, hence *monopole.* Dirac predicted the existence of elementary monopoles years ago, but none have so far been observed in the laboratory or in the cosmos. The fact that topological defects have not yet been observed poses a problem for theorists, as if geological theory predicted a Mt. Everest and Grand Canyon in the middle of Kansas.

Finally, in our list of problems, the standard Big Bang scenario gave no indication of how the lumpy structure of galaxies and stars could have formed out of the original hot homogeneous mixture of particles and radiation. This is called the *Inhomogeneity Problem:* The thermal fluctuations that existed at early times in the Big Bang should have resulted in a universe far more lumpy than the one we have.

None of these problems suggested that the basic idea of the Big Bang was wrong and should be discarded. The Big Bang theory was extremely successful in predicting the microwave background and in quantitatively accounting for the primordial production of helium, deuterium, and lithium. The notion that the universe is an exploding gas of particles and radiation is almost certainly correct; it was just incomplete as it stood in the late 1970s.

A breakthrough toward completing the picture was made in 1980 with the introduction of the concept of the *Inflationary Universe.* The credit is usually given to a young elementary particle theorist named Alan Guth, then working at Massachusetts Institute of Technology, who coined the catchy phrase and gave the most complete early account of the idea (Guth, 1973). As is often the case, however, others had glimpsed the idea earlier and now find their original work rarely cited.

The theory of the inflationary universe proposes that at about 10^{-35} second after the beginning of time the universe experienced a period of extraordinarily rapid expansion during which, within a tiny fraction of a second, the universe increased in volume by a factor of 10^{50} or more.

Let us see how inflation works. Consider when the universe was in a highly symmetric state, with a temperature of about 10^{27} degrees. Like the compass needles inside a hot piece of iron, any fields present would have had random directions that canceled each other, giving a net field of zero. As the universe cooled, these fields would freeze in place in a particular direction as the universe underwent a phase transition analogous to a ferromagnet. As in the case of a ferromagnet, the resulting broken symmetric phase is energetically favored when the temperature cools below a critical value. That is, the energy density is lower than it was in the original symmetric state and, all other things being equal, the system prefers to find itself in this state of broken symmetry.

Suppose the symmetric state is empty of particles—that is, a vacuum. Even though the field vectors add up to zero, each has energy, so the vacuum is not a total void—though empty of matter, it contains energy. This situation is called a *false vacuum* (fig. 9.4). Most significantly, the

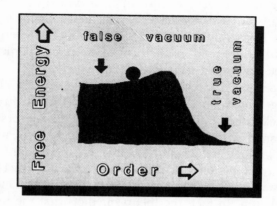

Fig. 9.4. As in the previous figure, we represent a system by a ball. Here it is shown temporarily trapped in a "false vacuum" state of higher energy because of a barrier near the edge. In the original version of the inflationary universe model, the exponential expansion occurs during the time that the universe remains trapped in the false vacuum, until quantum fluctuations send it over the barrier and down the hill to the "true vacuum" state of minimum energy and higher order.

fact that a vacuum can have nonzero energy is implied by the cosmological constant in Einstein's General Theory of Relativity.

Recall that Einstein had discovered that the equations of General Relativity allow for an arbitrary cosmological term that corresponds to an additional force beyond normal gravity that can be repulsive or attractive. This force is present even in the absence of matter or radiation, that is, in a vacuum. In this case normal gravity, which requires the presence of mass, is also absent. Einstein thought that this vacuum force must be repulsive to balance the attraction of gravity and stabilize the universe. Only later, when Hubble observed that the universe was expanding, did Einstein happily drop the term. Now, after years of disrepute, the cosmological term has reappeared.

In the inflationary theory of the early universe, the cosmological term comes into play not to stabilize the universe, but to expand it far more rapidly than in the current Hubble expansion. For the last 10 or 20 billion years, the galaxies have been moving apart with the velocity obtained in the initial explosion. However, during a preliminary inflationary era lasting about 10^{-30} second, the universe increased in size by 50 or more orders of magnitude under the action of the free energy of the false vacuum, as expressed by the cosmological term. When the inflationary period ended with a phase transition to the true vacuum, the free energy of the false vacuum was then released in the form of enormous particle production, producing the matter and radiation of the universe.

How does the concept of inflation solve the problems with the standard Big Bang model? First, there is no Horizon Problem. At the time the background radiation was created, the universe would have been much smaller than in the standard model. The various pieces of the universe at that time would be well within causal contact.

The Flatness Problem is also solved: according to Guth the rapid inflation flattens the universe just as the surface of a balloon is flattened by rapid inflation. In fact, the theory predicts that the universe should now be exactly flat, with the mass density precisely equal to its critical value—just enough for the universe to continue to expand indefinitely.

If future observations should disclose that the universe is not flat, the inflationary theory would be invalidated. This, incidentally, is a point in its favor. The theory is not simply speculation but makes a prediction that can be rigorously tested by future observations. A theory that is falsifiable is a good theory. Inflation meets this condition.

Remarkably, the Monopole and Domain Wall Problem is also solved by inflation. Defects are still formed, but our universe lies well within the walls of one domain.

The Inhomogeneity Problem—explaining the density fluctuations that eventually coagulate into the stars and galaxies—also appears solvable, although not yet completely settled, within the framework of inflation. Recall that, in the old Big Bang model, the fluctuations are too large to give our rather smooth universe. In the inflationary picture, the universe is much smaller when the fluctuations occur, so these fluctuations also are much smaller. Furthermore, inflation seems to give the qualitative features that are required for the inhomogeneities to be consistent with observations: about the right magnitude and equal fluctuations for all the various scales of distance involved. Unfortunately, cosmologists have hit some serious snags in developing the details of this issue, so we must be careful to keep our minds open to the possibility that the whole idea can still come crashing down at some future date.

As often happens, the concept of inflation solved old problems while introducing new ones not previously recognized as problems. In the original form proposed by Guth, inflation had a flaw; to understand it, we can rely on our experience with phase transitions in ordinary matter. When water is rapidly cooled below its freezing point, it does not immediately freeze into a solid block of ice. If the water is completely pure, it can remain indefinitely in a *supercooled* liquid state. A barrier exists between water molecules that prevents them from sticking together to form ice. To get over that barrier requires the existence of impurities in the liquid, normally present in even the most highly purified water, on which crystals of ice can nucleate.

In a similar fashion, Guth had proposed that our universe remained trapped in a supercooled symmetric state, even as the temperature cooled below the critical point for the phase transition to the condition of broken symmetry we now enjoy. Inflation ended when the universe eventually escaped into the broken phase. Like the freezing of water, the supercooled phase has to overcome a barrier to reach the broken symmetric state. This was assumed to happen by *tunneling,* a common and well-understood occurrence in quantum systems: a particle is able to tunnel through a barrier that is impenetrable in ordinary classical physics and common experience. Tunneling happens when the natural fluctuations in the position and energy of a particle, brought about by the Uncertainty Principle, allow it to jump the barrier.

The problem is that the nucleation process produces bubbles of broken symmetric phase surrounded by regions of symmetric phase. These surrounding regions continue to inflate exponentially, and the bubbles are never allowed to coalesce but rather are driven farther and farther apart. This produces a universe that is far more inhomogeneous than the one we observe.

The answer to this problem has come to be referred to as the *New Inflationary Universe,* proposed independently in 1982 by Andrei Linde in the Soviet Union (Linde, 1982) and Andreas Albrecht and Paul Steinhardt in the United States (Albrecht and Steinhardt, 1982). Instead of the high barrier that holds in the supercooled symmetric phase during inflation, this model says that no barrier exists. Rather, what is called a *slow rollover transition* occurs.

The situation can be likened to a ball on top of a mountain peak. In our case the universe is like the ball. The mountain top corresponds to the higher energy symmetric state; at the bottom is a deep valley containing the lower energy state of broken symmetry that we desire

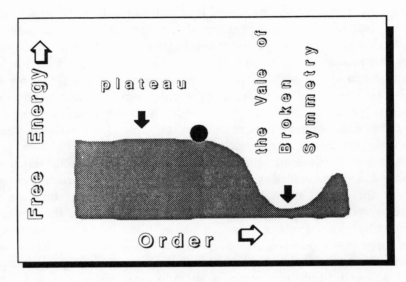

Fig. 9.5. In the new inflationary universe model, the universe remains on a long plateau of higher symmetry and lower order while it expands from this false vacuum state. Then it drops into the Vale of Broken Symmetry in a slow rollover transition. Particles are produced as the universe oscillates about the minimum.

to reach. This is the *Vale of Broken Symmetry* (fig. 9.5). In the old inflationary picture, the ball is in a little gully at the mountain top and has to surmount the gully wall before it can roll down the side. Thus it rolls back and forth in the supercooled state as the universe expands, awaiting a quantum fluctuation big enough to send it over the top and down the side to its true energy minimum.

In the slow rollover picture, the gully does not exist, or is very small, whereas the mountain top is very flat. A slight push in one direction will cause the ball eventually to fall down the side, but there will be a sufficiently "long" period (10^{-30} second!) while it remains at the top in the false vacuum state. During this period, the universe inflates.

Once the universe falls down the hill, it reaches the valley of minimum energy. Here it will eventually settle, but first it will roll back and forth like a marble dropped into a bowl. As the fields oscillate about their equilibrium value, they will radiate particles, just as an oscillating electromagnetic field radiates photons. The energy the universe had on top of the mountain must be dissipated and is very efficiently transformed into relativistic particles—photons, quarks, and leptons, or their precursors—which eventually form the atoms and radiation of the universe.

At this writing the details of the mathematical models to bring about the slow rollover phase transition have not been developed. Perhaps this will turn out to be impossible. However, such phase transitions are known to occur in matter, notably in a magnet where no barrier prevents the freezing in place of the magnetic domains. Thus, we currently can anticipate that a model will ultimately be found.

Many of the details of the inflationary picture presented here likely will change with future developments. Nonetheless, the universe is evidently expanding from an initial condition of maximum entropy. Further, order can be spontaneously generated while the universe expands. These solidly based statements allow us to anticipate that, whatever form future theoretical developments take, these developments will not change our basic conclusions: the order we see is the result of the accidental ways the original symmetries were broken during one or more phase transitions in the early universe.

I picture the origin of the universe as follows: in the beginning there was a void more empty than a perfect vacuum, empty not only of particles and fields but of space and time as well. It had perfect symmetry and zero energy. It was as much nothing as nothing can be. A fluctuation in that void then occurred, generating our universe and perhaps countless others very different from ours. But a question still remains that requires a response: How can something come from nothing?

Our traditions tell us that some supernatural process is necessary to produce something from nothing. But we have seen that tradition has been wrong before, telling us that the earth is flat and at rest and that space, time, and mass are absolute. Unfortunately, our language does not yet contain adequate words to describe something without space and time. Even discussions of so-called spiritual phenomena are phrased in terms of space and time—which, to me, places them back in the realm of physical phenomena and not that of the spiritual. Our grammatical structures require the use of verbs with tense, based on the assumed existence of past, present, and future. We are forced to use terms such as *where, when,* or *then.* So in discussing the origin of the universe, whatever we say with these inapplicable words may not always be a fair or accurate rendition of the situation. Even our mathematics is not yet quite up to the task. However, it does help us on one point.

Over the centuries we have developed the concept of *zero* and the related *negative,* which is the quantity added to another quantity to give zero. This has proved useful in many aspects of our everyday lives, as well as in science. We keep accounts with positive (credit) and negative (debit) columns, which we are obliged to balance. We classify many human actions and motives as positive or negative, and talk about political decision making as a zero-sum game. So if the universe was originally nothing, it could now be something—as long as the positives and negatives still add up to zero, at least at the Planck Time when there was the perfect symmetry of zero.

Science is replete with positives and negatives in its description of phenomena at all scales; for example, the opposite charges and forces in an atom cancel each other. While we normally think of the kinetic energy of motion of a body as positive, the potential energy that a body can have in the presence of other bodies can carry either sign. In particular, potential energy carries a minus sign when it results from an attractive force between bodies. For example, a body on the surface of the earth has a negative potential energy; a body coming in from space falls down this *potential well.*

If a body is tossed in the air, it will normally return to earth because its kinetic energy is insufficient to overcome the large negative potential energy the body had by virtue of being in the large gravitational field of earth. The total energy—the sum of the kinetic and potential energies— is negative. If the body is given a velocity of 11 kilometers per second, its kinetic energy will exactly balance its potential energy, so that the total energy is zero, and the body will be able to just barely escape into space. This particular velocity is called the escape velocity (Incidentally, note that the escape velocity from the earth is greater than the seven kilometers per second or so necessary to put a satellite into near-earth orbit; in that case the satellite is still bound to earth with negative total energy.)

We have already remarked that the universe appears to be balanced between open and closed. This is precisely the condition of zero total energy, with the total rest and kinetic energy of all the bodies in the universe canceling the total potential energy. Since the universe has zero total energy, no energy was required in its production.

To demonstrate this important point quantitatively, let us consider the spontaneous generation of two particles of mass m. From Newton's

Law of Universal Gravitation, they will have a gravitational potential energy given by $-Gm^2/r$, where G is Newton's constant and r is the distance between the centers of the particles. For simplicity, assume the particles are produced at rest, so that the kinetic energy is zero. The rest energy of each particle is equal to mc^2, from Special Relativity. Thus, the total energy of the system is $E = 2mc^2 - Gm^2/r$. Clearly, the total energy will be zero for $r = Gm/2c^2$. In other words, it is always possible to spontaneously generate two particles of mass m from zero energy, as long as they are separated by the required distance when they appear. For the known elementary particles, the value of r is smaller than the smallest possible distance, the Planck Length of 10^{-33} centimeter, so this does not happen except "virtually" for small times allowed by the Uncertainty Principle. Particles of the Planck Mass, 10^{-19} GeV, however, will be produced at distances comparable to the Planck Length.

Although the energy of the universe appears to be in balance, the type of matter does not. Normal matter outnumbers antimatter by a factor of about a billion. However, we have already seen how the currently observed excess of matter over antimatter is neatly explained as one of the broken symmetries in an early universe phase transition. This implies that the total baryon number of the universe—the sum of the particles and antiparticles—was zero at the beginning and only became nonzero after this phase transition occurred.

The total electric charge of the universe is almost certainly zero. In fact every known quantity in the universe normally conserved in physical processes is consistent with a result of zero when summed over all bodies. Even the mass of the universe is included in this energy balance, since mass and energy are equivalent.

If we cannot use space and time in our description of events prior to the Planck Time, we may still be able to use quantum mechanics. Early in the development of quantum mechanics physicists recognized that there were phenomena, such as spin, that defied description in terms of conventional notions of space and time. As a result, much of quantum theory was formulated independent of the need to represent all physical phenomena exclusively in terms of space and time. For example, one of the basic tenets of quantum mechanics is independent of space and time: the *Principle of Superposition,* in which the state of a system is said to be in reality a combination, or superposition, of all the possible states of the system. Thus a system may be in one state, but it may

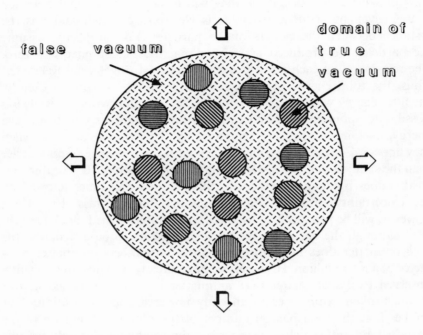

Fig. 9.6. The domains of order (true vacuum) form in a sea of disorder (false vacuum). The latter have a higher energy density, which causes them to expand more rapidly. Each domain can undergo further phase transitions, becoming a separate universe with different physical laws than the others.

also be in another, with the particular outcome of a measurement depending on chance. For example, a photon can be thought of as being an electron-positron pair, or a quark-antiquark pair, or any particle-antiparticle pair. The effects of this ambivalence are real and measurable in atomic and particle physics. The vacuum is found to be electrically polarized, having some nonzero probability for containing equal and opposite charges separated from one another for short periods of time. The effect of this polarization has been measured to great precision and calculated to matching precision in quantum electrodynamics.

In the same way, a totally empty state of the universe will be a superposition of all the possible states in which the energy, charge, baryon number, and all other physical quantities of this type add up to zero. These states will have the perfect symmetry of the void. This includes a universe with particles whose total kinetic energy is exactly balanced by a negative potential energy. It also includes a universe in which a large number of randomly oriented fields all add to zero.

So the totally empty void has some probability of being found in any one of these zero-sum states. We can imagine that one of these states spontaneously occurred with different domains of positive and negative energy density (fig. 9.6). The positive density domains inflated while the negative density ones deflated. The deflating domains collapsed, while the inflating domains expanded beyond the Planck scale. As they expanded they cooled, and additional domains of broken symmetry phase formed, surrounded by a continually inflating symmetric phase. One of these subdomains of broken symmetry became our universe, with its particular set of rules or order brought about by the random selection of the field vectors that happened to freeze in. Other expanding domains became other universes, with totally different natural laws and forever beyond our ken.

10

We Will Become God

But multitudinous atoms, swept along in their multitudinous courses through infinite time by mutual clashes and their own weight, have come together in every possible way and realized everything that could be formed by their combinations. So it comes about that a voyage of immense duration, in which they have experienced every variety of movement and conjunction, has at length brought together those whose sudden encounter normally forms the starting-point of substantial fabrics— earth and sea and sky and the races of living creatures.

Lucretius, *De Rerum Natura*

Most people find comfort in the notion of a universe set down according to plan. They assume this is the only way to find meaning in their individual lives, believing that meaning can only be provided from without. I personally prefer a world in which I have primary control over my own destiny, providing my own meaning by my own actions, for better or worse. Beyond this subjective preference for a world of individual control and responsibility, I think the notion of a grand design is one that cannot withstand the twin objective tests of logic and observation.

Once we assume there is a grand design, the disorganized reality we see around us becomes almost impossible to explain, unless it too is part of the plan. The presence of a plan implies that events occur in a preordained way. I suppose we can imagine a deity who sits there stirring the pot, but I do not think this is what most people mean by a Creator. The conventional notion holds that the Creator had some intent when He made the universe. The misery, disorder, and injustice of the world are dismissed as mysteries our minds are incapable of comprehending. However, if we do not assume the existence of a Creator right from the start—as unquestioned, self-evident truth—then the disorder in the world can much more reasonably be seen as evidence for lack of plan rather than for one purposefully hidden from our view.

In this book I have tried to describe what is beginning to develop as science's view of an accidental, unplanned origin of the universe. In doing away with the notion of a Creator, we fly in the face of some of the most sacred beliefs of the world's great religions. These beliefs go back thousands of years, so believers might argue that they have the proof of venerability on their side. After all, how could so many great thinkers over so many centuries be wrong? But recall that these same thinkers also believed that the sun revolved around the earth, and their belief did not make it so.

However, let us grant the premise that people who lived thousands of years ago could have grasped insights that are still valid today. The tradition of science and scientific method—which is leading us to opposite conclusions on the origin of the universe—is just as venerable, going back to times as early or earlier than the origins of all today's major religions, with the exception of Hinduism.

On the spring day in 585 B.C.E. when the sun was covered in Miletus, the Hebrews were on the road to exile in Babylonia after the destruction of Jerusalem by Nebuchadnezzar. At the same time, the Persian sage Zarathustra, or Zoroaster, was preaching a religion centered on a single god, Ahura-Mazda. Buddha would be born in India twenty years later. Confucius would be born in China about ten years after that. And, of course, Christianity and Islam would not appear for centuries.

So if men have held a mystical and magical view of the universe for many centuries, others living in those same centuries were beginning to glimpse the idea that not mysticism or magic, but the observations of our senses offer the surest path to truth. The accidents of politics and history, and perhaps some psychology not yet understood, caused the mystical view to gain wider acceptance and greater influence on human minds for millennia.

It is a convenient excuse for religions to call anything they cannot explain a mystery beyond human comprehension. Science is unwilling to make any assumptions about the limitations of the human mind in understanding the mysteries of the universe, including its origin and the source of the order that is observed. And these are no less scientific questions than the density of water. Why should religions insist they have a special insight on this question and tell science to confine itself to the more mundane issues—especially when all their previous attempts at explaining the world have been wrong?

Nevertheless, even among scientists the belief that the universe is obedient to a perfect natural law, whose secrets are gradually being unfolded as humankind progresses, is deeply ingrained by our Judeo-Christian heritage. The world obviously is not perfect. This discrepancy is no less difficult for a scientist holding this view to explain than for a priest of any of the mystical faiths.

The view that the universe was not the result of a conscious act challenges not only religious tenets but also many great scientific thinkers, including Newton and Einstein, who believed in a grand design to the universe. But even Einstein developed his Theory of General Relativity at a time when the stars were thought to reside at fixed positions in a firmament, and in that model of the universe, the principles of physics supported the notion of evolution from a more orderly state.

As discussed earlier, when the Second Law of Thermodynamics was discovered in the last century, scientists inferred that the universe must have begun in a state of high order and is now proceeding toward a state of ultimate disorder, called the *heat death*. They did not know, as we do today, about the constantly increasing maximum entropy of the expanding universe and the tiny fraction of that universe which constitutes the matter of the galaxies. We have seen that this leads to quite the opposite conclusion: The universe *could* have happened by accident, and arbitrary natural laws *could* result from the accidental way the original lawless symmetries were broken as our particular universe expanded and cooled during the first fraction of a second of its existence.

Some might argue that this does not *prove* that the universe is an accident, and they would be correct. But I would respond that it is not up to the skeptic to disprove concepts not justified by the data. Rather, the burden of proof lies with those who claim to see a pattern in nature beyond the expectation of blind chance. They hold the prime responsibility for convincingly demonstrating that more than imagination or wishful thinking leads to their conclusion. They must give convincing reasons to introduce more concepts than are needed to describe the existing data. The purpose of the preceding chapters has not been so much to explain how the universe actually came about, as to show that nothing known from observations about the physical universe forces us to conclude that it had to occur by plan.

If any single principle allows us to approach the conclusion that the universe did not occur by plan, freed from the fetters of our subjective psychological needs, it is the *cosmic perspective*. As much as we might individually wish it to be otherwise, an openminded view of the realities presented to our senses leads us firmly to the recognition that the earth and humankind are almost incomprehensibly negligible portions of reality. The earth is not the center of a universe created for human beings. The very matter that composes people, the earth, and the hundreds of billions of stars and galaxies is a tiny bit of dust left over when the universe exploded.

If the universe is an accident, so are the earth and humankind. And the evidence that this is so abounds, once we examine it without the arrogant preconception that we humans are perfect creations, pre-ordained to rule the universe. Life began when chemical reactions in the atmosphere and oceans of the early earth manufactured complex multi-atom molecules. That this can happen spontaneously has been amply demonstrated in the laboratory and is also evident in the cosmos, where organic compounds are found in the spaces between stars and in meteorites. When conditions of temperature and pressure are right, and energy sources exist, complex molecules will inevitably form by chance collisions and the simple rules of gravity and electromagnetism. Calculations which purport to show that life is too improbable to have occurred spontaneously are simply wrong. With millions of years under slowly changing environmental conditions, molecules can become more complex and can lead to the high level of localized organization we identify as life. As we have seen, the increasing organization of life can easily be accommodated within an overall entropy increase of the rest of the universe, specifically the photon background, as required by the Second Law of Thermodynamics.

If organized matter formed on the earth under the conditions there, we might expect that it has occurred, and will occur, elsewhere. So far we have found no evidence for life beyond earth in our own solar system, and searches for radio signals from extraterrestrial civilizations have so far proved negative. From what we know, life is a highly unlikely occurrence, so these early null results should not be surprising. Having only one example of life, we simply have insufficient data to compute with any confidence the probability of life on other worlds. Estimates range from many civilizations per galaxy to us as the only life anywhere, in any galaxy. Life, like the universe itself, could be simply an unlikely aberration.

Current searches for signals from other civilizations are based on the highly anthropocentric notion that such civilizations would communicate the way we do, with radio waves. I see no reason to assume this, or that other life forms are carbon-based as we are or necessarily similar in size, appendages, or sensory apparatus. Our solar system is in the boondocks of the galaxy, two thirds of the way out from the center. Perhaps civilizations exist in the denser central regions and communicate by sending neutrinos back and forth through the thick matter in between. No evidence exists that extraterrestrial beings, ancient or modern, have ever visited the earth, despite many claims in the popular literature. It still is possible, however, that the universe teems with civilizations. This is a great unanswered question.

Charles Darwin's discovery of the evolution of life from simple beginnings is now widely confirmed in principle, although many details are still being uncovered as science learns more about the mechanisms involved. Recently it has become evident that evolution is not smooth, as Darwin envisaged, but occurs in discrete jumps. This concept is called *punctuated evolution*. This realization has occurred in parallel with the recognition that the physical environment of the earth—and indeed the universe—is not as stable as once was believed, so species are often forced to adapt to drastically changing situations in order to survive. Many species fail to adapt. Far more species have become extinct with the passage of time than exist today, and species unable to adjust are continually dying out.

Evidence is accumulating that every 26 million years the earth experiences some major upheaval that results in wholesale extinction of species. The dinosaurs are believed to be one of the casualties of an event about 65 million years ago; perhaps 75 percent of all species were obliterated at that time. The timing of these events is evidently so precise that any explanation is almost certainly astronomical. The current theory, supported by geologic evidence, is that large comets or meteors hit the earth at these regular intervals, raising an enormous dust cloud that blankets the sun for several years. This is a natural version of the *nuclear winter* scenario that is expected to result from any nuclear war. In fact, the prospects for nuclear winter were discovered by astronomers and other scientists studying the problem of the earth's response to a collision with a comet.

If large extraterrestrial bodies hit the earth on a regular basis, some periodic event in the cosmos must trigger an excess far above the normal collision rate of comets or meteors. One theory is that the sun has a dark companion star, called *Nemesis*, that regularly passes through the *Oort cloud* of billions of comets within one light year from the sun. The cometary orbits are disturbed sufficiently to cause an increased number to fall toward the sun, raising the likelihood that one will strike the earth. The last destructive event occurred about 11 million years ago, so we will have to wait another 15 million years to fully test the theory. In the meantime, astronomers are looking for Nemesis.

While the large-scale destruction of life may seem to be a terrible event, it has benefits that are crucial to the progress of life toward higher levels of development. In a stable environment, living things naturally develop greater specialization to that environment and eventually fill all the niches in the ecosystem. Greater specialization better equips them to find food and cope with their normal enemies, but in the process they may lose their ability to adapt to changes. A sudden change in the environment thus performs the immensely useful task of ridding the ecosystems of these highly specialized dead-end species, allowing room for more adaptable species to survive and expand.

While this great filtering out of overspecialized species may occur on a regular basis, it is pretty much accidental whether a comet will hit the earth with enough force to produce large-scale damage. Other more frequent cosmic events are causing accidental but continual damage to all living creatures, and thereby contributing to the changes that are necessary for life to evolve. Cosmic radiation, mostly protons, constantly bombards the earth. Muons produced by these protons in the atmosphere reach the surface and pass through all living things at the rate of about one particle per square centimeter per minute. These muons will occasionally strike an atom in a key strand of genetic DNA within a reproductive cell, which can result in a mutation, permanently changing some characteristic in an offspring.

In most cases the mutation caused by cosmic radiation or other natural radiation in the earth, or by many other nonnuclear sources, is minor. In a few cases it may be major and likely harmful to the recipient. In a very small fraction of cases, the change may be for the better, providing a slightly sharper tooth to enable the animal to eat some newly introduced food item, or lightening its exterior color to better blend in with a snowier landscape brought about by a cooling climate. So while the individual plant or animal, or even a whole species, probably suffers from catastrophic changes, life as a whole often benefits as the more adaptive features survive. In that way, a rather weak and slow animal called *Homo sapiens* has survived thus far and come to dominate the earth, by being among the most adaptive of natural earthly organisms, although perhaps not as adaptive as the cockroach or rat. If man is ever forced out of his ecological niche, it will have to be by an even more adaptive species. Hopefully we will be able to hold off the cockroaches and rats.

Until recently there has been a great misunderstanding about life and its relation to the environment of earth. It previously had been assumed that living things exist in a sort of harmony or equilibrium with their environment. This was called the balance of nature and was taken as evidence for the grand design that so many so desperately want to find. But in the last few years the study of ecology has uncovered the fact that the elaborate feedback system that is supposed to maintain every organism in a symbiotic equilibrium is hardly ever achieved; life exists under constantly changing circumstances. While there certainly is feedback in both the physical and biological systems of the earth, natural events and human activity constantly upset the balance in ways that are sometimes regular and sometimes random. Nature may be balanced at any given time, but the balance is a precarious one, and a slight shove one way or another can cause the system to change and oscillate, until the next precarious balance is achieved. The balances that do exist in nature are far from eternal.

Another common misconception is that living things are perfectly adapted to their environments, with just those appropriate characteristics which enable them to survive most efficiently as a species. In fact, many of the characteristics of plants and animals have little or even negative survival value. They were just accidents and were maintained if they resulted in little loss of survivability. Also, given the fact that the environment can unexpectedly change, an animal is fortunate to have a few characteristics beyond those it needs for its current circumstances.

So we have two arms, two legs, two eyes, and ten fingers by chance, and they work well for us. Certainly another eye in the back of the head would be even better, and our toes do not seem to serve much purpose anymore. If any one of us, much less a deity, set out to design a human being, he undoubtedly could do a far better job than chance.

Indeed, humankind is in the process of creating a greatly improved version of itself, not by the painfully slow and largely random process of biological evolution but by the rapid and guided advance of technology. This new form of "life" I will call, for historical reasons, the *computer*. But please understand that by this I mean something much more than today's personal computers or even supercomputers.

Of course computers are not alive in the traditional narrow use of the term, as applied to the naturally evolved carbon-based plants and animals of earth. But computers have many of the characteristics of living things, as well as many more of their own—and the term *life* is an arbitrary designation anyway. There is no fundamental difference between organic and inorganic matter, except the conventional one used by chemists, which defines *organic* to mean containing carbon. And there is no special *vital force* in living things that makes possible a distinction. It's all in the atoms. The main characteristics we associate with life are a high level of organization and self-motivation. Computers certainly have the first and are on the way toward achieving the second.

Instead of basing this new life form on the carbon atom, which just happens to form the framework of earthly biological life, a more favorable element, silicon, is currently in primary use. Other materials will surely be applied as they become available and are proven to be superior. That is the beauty and strength of this new artificial life form. Computers are far more flexible and adaptable than any natural life form. They are not locked into some specified shape or fixed number of appendages; rather they take on the shapes, appendages, and other qualities needed to serve a purpose. Computers are not restricted to sensing the narrow bandwidths of visible light or audible sound. They can see the ultraviolet and infrared and hear supersonic sounds or solar neutrinos. Computers do not need to come in individual units, although this is sometimes useful, as in the case of the home computer. They can be networked with other computers, providing communication and contact between units far more efficiently than is possible between human beings.

Computers, in their robotic manifestations, are already far superior to humans in most manipulative tasks. Right from the start, they were superior in most calculational jobs—once the proud preserve of the human mind. This is the reason we first called them computers. But do they think? Are they intelligent?

Computers obviously do not think quite like humans, and I do not claim that the computer of today is necessarily a valid model of the human brain. But computers process data and make decisions based on that data, which is all that the human brain does under the label of thinking. A prime area of study today is *artificial intelligence* (AI), and its practitioners harbor few doubts that someday computers will be made to do all the operations normally associated with human intelligence, and many more. If the intelligence that results is not strictly human, that is not to say that it will necessarily be inferior. Perhaps artificial intelligence will be superior, with characteristics and capabilities the human mind can not even imagine.

Computers are certainly more logical, besides being computationally faster, than the human brain. And they can do creative work as well; witness computer-generated music and art. Computers are still guided by human programmers in all these tasks, but they are beginning to perform many of their operations free of rigid human control. Computers no longer stupidly follow only those steps laid out by the programmer; they are capable of making choices unanticipated by the programmer. For example, they can consistently defeat their own programmers in chess.

Many writers, in a desperate search for qualities that will justify their prejudice that human beings are somehow different from and superior to machines, have found examples of problems the brain can solve and computers currently cannot. However, we should not judge the potential capabilities of computers by those in current use. After all, they are still a new "species," having been in existence for perhaps half a century and reproducing only four or five generations. Homo sapiens has already reproduced thousands of generations.

If computers in their current early forms do not possess some conventional life-like characteristics, such as the ability to heal or reproduce themselves, there is no doubt that they will be able to do this in the near future. Already computers are used to design other computers, and to help fix the ones that fail. Human intervention in these operations is becoming less important with each new improvement.

Computers still have trouble with certain tasks, such as human vocal language recognition. However, computers are not human, and their own languages are in many ways superior to cumbersome, ambiguous, randomly developed human tongues. The problem with human language recognition is on the human side, not the computer side. Our spoken sounds and written words are too illogical and ambiguous. If there is anything we do that computers cannot, be patient. In time they will do it better, if it is worth doing at all. A good example of this is the exploration of space. We spend millions of dollars putting men and women into space where they perform few significant tasks that cannot be done better by computers. The justification for this huge expenditure is not scientific, but political and emotional. We have had politicians in space, and the first teacher in space tragically lost her life. I would love to walk on Mars; but do I really *have* to walk on Mars? Computers, which do not need oxygen and water, much less food and energy, will gather far more data than I ever could in my cumbersome spacesuit. Computers are subject to damage from radiation or space debris, but far less so than people. Our bodies evolved within the very narrow environmental confines of the surface of the earth. We are cold or hot under changes in temperature that are a tiny fraction of the ranges occurring in space. We require an equally small range of atmospheric pressures, and cannot exist in a weightless condition for long. All these deficiencies can be overcome in space with expensive life-support systems, but to what purpose other than the satisfaction of our own egos? Why is it so important that a human being be out there, as long as the data come back?

And what about the future exploration of the universe beyond the solar system? We humans do not live long enough for travel between the stars to be even thinkable in the forseeable future. In the time it will take us to develop the technology to travel to the stars, we will likely have been moved out of our ecological niche by our own creations, the computers.

Future computers will not only be superior to people in every task, mental or physical, but will also be immortal. An individual unit may wear out and become inoperative, of course, but its "soul"—the memory banks of information it holds—can be transferred to another unit when the first has had its day. *In this way we also can become immortal.* It should be possible in the future to save the accumulated knowledge of an individual human being when he or she dies. Perhaps even those thoughts which constitute consciousness will also be saved, and the collective thoughts of all human beings will be continued in the memory banks of computers. After all, these thoughts in the human brain are electrical impulses that are no different in principle from those circulating inside a computer memory bank.

Most people, with our deeply embedded anthropocentric traditions, may be expected to have very negative initial reactions to these ideas. But think about the question objectively. Given our own severely limited physical bodies, we can never hope to live longer than a century or to explore much beyond the confines of earth. Imagine, however, being part of a collective consciousness of all humankind and computerkind, with mental powers and sensory inputs infinitely superior to the ones we now possess. We would be able to explore the universe, using our superior sensors to "see" lights and "hear" sounds over spectral bands that far exceed those of our current eyes and ears. With our infinitely expanded mental capacity, we would be able to think thoughts, to enjoy pleasures of beauty and intellect, beyond our wildest dreams and fantasies.

If computer thinking now is cold and logical, what is to prevent it from eventually encompassing all those other aspects we associate with thinking? If the computer is "just a machine," so is the human brain, albeit a quantum mechanical one. If the human brain can think thoughts that are emotional rather than logical, and if there is a benefit to this as there indeed seems to be in the enhancement of creativity and social good, then this future machine consciousness will be capable of these types of thoughts, and new types not yet imagined. Perhaps, as part of this new consciousness, we will become God. If we do, it will have been by accident.

Glossary

Anthropic Principle. The universe has the structure that led to human existence because, if it were otherwise, we would not be here to comment upon it.

Antimatter. Matter composed of particles that have electric charge with opposite sign to the particles of normal matter.

Artificial Intelligence. Computer intelligence.

Avogadro's Number. 6.022×10^{23}. The number of nucleons (protons and neutrons) in 1 gram of matter.

Baryon. A particle that is composed of three quarks, such as the proton or neutron.

Baryon Asymmetry Problem. The fact that the universe contains a billion times as many baryons (ordinary matter) as antibaryons (antimatter), when they would be expected to have been produced in equal quantities in the early universe.

Baryon Number. The number of baryons in a system. Antibaryons count negatively.

Big Bang. The explosion of the universe. Usually refers to the period after inflation.

Binding Energy. The energy that holds a molecule, atom, or atomic nucleus together.

Black Body Radiation. Electromagnetic radiation from a body with a smooth wavelength spectrum that depends only on the temperature of the body.

Black Hole. A body whose mass is contained within such a small volume that gravity prevents the escape of anything, including light.

Boson. Any particle with integer spin, such as the photon, weak boson, gluon, graviton, or composite objects such as helium nuclei.

Broken Symmetry. A slight deviation from perfect symmetry. The process by which physical laws are spontaneously generated.

Brownian Motion. The random motion of macroscopic particles that results from their being bombarded by molecules.

Charge Conjugation. A mathematical operation in which a particle is changed into an antiparticle. Also particle-antiparticle conjugation.

Charm. The quantum number carried by *c* quarks.

Classical Physics. Physics before the quantum revolution. Technically still includes Einstein's theories of relativity, although this distinction is not always followed.

Closed Universe. Model in which the average density of the universe is so high that the universe will eventually stop expanding and contract back to a point.

Color. The equivalent of electric charge in the strong nuclear interaction. The property carried by gluons as they are exchanged between particles.

Compton Scattering. The scattering of photons by electrons or other particles.

Conservation Principles. Quantities such as energy, momentum, angular momentum, and charge remain constant in chemical, nuclear, and particle reactions.

Copernican Principle. The earth is not the center of the universe.

Correspondence Principle. Quantum mechanics must give the same results as classical mechanics when applied in the domain where classical mechanics is valid.

Cosmological Principle. The universe looks the same—that is, is described by the same laws—at every place and every time.

Cosmological Term. A term that appears in Einstein's General Theory of Relativity and allows for the possibility of additional attractive or repulsive gravitational forces. Occurs even in the absence of matter or radiation.

Curie Point. The temperature below which iron becomes magnetic.

Dark Matter. The still unobserved main component of the universe.

Decay Product. Any of the particles that are produced when an atomic nucleus or particle decays.

Diffraction. The bending of sound or light around corners.

Doppler Effect. The increase or decrease in frequency observed from a source of sound or light, when that source is moving toward or away from the observer.

Domain Wall. The boundary between two different phases, such as liquid and gas.

Electroweak Unification. The unification of electricity and magnetism with the weak nuclear force.

Elementary Particle. Any particle that is not known to be composed of smaller particles.

Entropy. The measure of disorder or lack of information.

Escape Velocity. The minimum velocity needed to escape from the gravitational pull of a body.

Exchange Force. A quantum mechanical force that has no classical analogue but plays a major role in holding atoms together into molecules.

False Vacuum. The situation in which a volume contains no particles or radiation and yet is not the state of minimum energy.

Fermion. Any particle with half-integer spin, such as the electron, quark, proton, or neutron.

Fitzgerald-Lorentz Contraction. A moving body is observed to contract in length in the direction of its motion.

Flatness Problem. The universe is very close to being Euclidean. Since any nonzero curvature changes with time, the standard Big Bang model cannot explain this.

Gamma Ray. A photon with an energy higher than that of an X-ray photon, characteristic of photons in nuclear and elementary particle reactions.

Gauge Symmetry. A general class of symmetry principles that are the key ingredient in unification schemes.

Generalized Copernican Principle. There is no center of the universe, no special point in space or time.

General Theory of Relativity. Einstein's non-Euclidean geometrical theory of gravity in which gravitational effects result from the curvature of space near massive bodies.

Geodesic. The path that would be taken by a light ray in moving from one point to another in a curved space. Also, a great circle on the earth.

Gluon. The quantum of the strong nuclear force. Exchanged between quarks to produce the force.

Grand Design. A plan to the universe that did not occur spontaneously.

Gravitino. The supersymmetric partner of the graviton. Not yet observed.

Graviton. The quantum of gravity. Exchanged to produce the gravitational force.

Hadron. Any strongly interacting particle.

Heat Death. The theory that the universe must eventually become completely disordered.

Heisenberg Uncertainty Principle. The uncertainty in the measurement of the position of a body times the uncertainty in the measurement of its momentum is greater than or equal to Planck's constant. It is impossible to measure both position and momentum with infinite precision. The same principle applies to measurements of energy and time.

Hilbert Space. An abstract mathematical space used to describe quantum mechanical states.

Horizon Problem. The temperature of the microwave background is the same for regions that could not have ever been in causal contact in the standard (pre-inflationary) Big Bang model.

Hubble Factor. H = 15-30 *kilometers per second per million light-years.* The factor by which one multiplies the distance of a galaxy to get its speed of recession, on average.

Hubble's Law. The distance to a galaxy is proportional to its speed of recession, on average.

Inflationary Universe. The theory that the universe underwent exponential expansion during the first fraction of a second, before the linear expansion of the Big Bang. Solves many of the problems with the standard Big Bang model.

Inhomogeneity Problem. The visible universe is less homogeneous than can be understood in the conventional Big Bang model. How did galaxies form?

Internal Space. Additional dimensions, besides the four of space and time, that are used to describe other particle degrees of freedom such as spin, charge, and baryon number. They may be spatial dimensions that are curled up at such small distances as not to be directly observable.

Internal Symmetries. Symmetry principles that apply in internal space.

K-Meson (Kaon). A strange particle of zero spin and about half the mass of a proton. Contains an *s* quark.

Left-Right (Mirror) Symmetry. The laws of physics are the same for mirror images of natural events as for the original events.

Lepton. A half-integer spin elementary particle such as the electron, neutrino, or muon. Does not interact strongly.

Leptoquark. A generic term that refers to both quarks and leptons in Grand Unified Theories.

Magnetic Monopole. A particle containing the magnetic equivalent to electric charge. Not yet observed but expected to exist.

Matrix Mechanics. Heisenberg's original form of quantum mechanics. Observable quantities are represented by matrices, *i.e.,* tables of numbers, rather than single numbers.

Mechanical Universe. A universe that is like a machine or clock, operating according to fixed natural law with everything that happens predetermined. Also known as the Clockwork Universe.

Meson. Class of integer spin particles such as the pion or kaon. Composed of a quark and antiquark.

Mesotron. Early name for the muon, now obsolete. Not to be confused with meson.

Minimal SU(5). The simplest Grand Unification Scheme; invalidated by the nonobservation of proton decay at the level predicted.

Monopole and Domain Wall Problem. Too many magnetic monopoles and other topological defects are produced in the early universe in Grand Unified Theories.

Muon. An elementary particle that is like the electron except that it is 200 times heavier. Not a component of normal matter but the most common component in cosmic rays at sea level.

Neutrino. An elementary particle of half-integer spin, zero charge, and zero or very low mass. Produced in the beta decay of neutrons.

Neutrino-Dominated Universe. The model in which neutrinos with mass constitute the dark matter of the universe.

Neutron Star. The extremely compact remnant of a star that has exploded as a supernova. Composed of neutrons and having the density of nuclear matter.

New Inflationary Universe. A modification of the original inflationary universe model that contains a slow rollover phase transition.

Non-Euclidean Geometry. Curved space geometries, as used in the General Theory of Relativity.

Occam's Razor. The principle that one should introduce no more hypotheses than are necessary to explain the data.

Omega-minus. A particle composed of three *s* quarks that was predicted by *SU(3)* and discovered shortly thereafter.

Open Universe. The situation in which the average density of the universe is low enough that the universe keeps expanding forever.

Pair Production/Annihilation. The creation or destruction of a particle-antiparticle pair, such as electron-positron.

Parity. The mathematical operation of changing the handedness of a physical system.

Parity Violation. The observation that the mirror image of a natural process is not always fundamentally equivalent to the original process.

Particle-Antiparticle Conjugation. See Charge Conjugation.

Photoelectric Effect. The emission of electrons from a material when it is illuminated with light.

Photino. The supersymmetric partner of the photon. Not yet observed.

Photon. The particle of light. Also, the quantum of the electromagnetic force, whose exchange is responsible for that force.

Pion. The lightest meson. Produced copiously in high energy particle interactions.

Planck Length. Smallest possible distance: 1.6×10^{-33} centimeter.

Planck Mass. Smallest mass black hole: 1.2×10^{-19} *GeV* $= 2.2 \times 10^{-5}$ gram.

Planck's Constant. $h = 4.14 \times 10^{-15}$ electron-volt-seconds. The key constant in quantum mechanics that relates energy to frequency.

Planck Time. Smallest possible time: 5.3×10^{-44} second.

Positron. Antielectron. Positively charged.

Preon. Generic name for constituents of quarks and leptons, if there are any. Not yet observed.

Principle of Equivalence. The equivalence of inertial and gravitational mass. The idea that acceleration and gravity are indistinguishable. The basic assumption of the General Theory of Relativity.

Principle of Galilean Relativity. Velocity is relative. There is no difference between being in motion at constant velocity and being at rest.

Principle of Superposition. In quantum mechanics, a state is equivalent to a combination of all the other possible states with the same quantum numbers.

Quantization Condition. Bohr's principle that values of angular momentum only occur as integrals of Planck's constant divided by 2π.

Quantum. A discrete unit. Also used to refer to particles such as the photon, weak bosons, gluon, and graviton that are exchanged to produce the various forces.

Quantum Chromodynamics (QCD). The quantum theory of the strong nuclear force.

Quantum Electrodynamics (QED). The quantum theory of electromagnetism.

Quark. Elementary particle which is the constituent of hadrons, such as the proton. Types: *u,d,c,s,b,* and *t.*

Quasar (Quasistellar Object). Distant galaxies that radiate far more energy than normal galaxies. Thought to be galaxies in the early stage of evolution.

Red Shift. The shifting of the wavelength of light toward larger values that results from their motion away from the observer.

Rotational Symmetry. The laws of physics do not depend on the orientation of the system.

Schwarzchild Radius. For a spherical body of a given mass, the radius below which, if all the mass is concentrated, the body becomes a black hole.

Second Law of Thermodynamics. The total entropy (disorder) of the universe stays the same or increases in any physical process.

Selectron. The supersymmetric partner of the electron. Not yet observed.

Slow Rollover Transition. A phase transition that does not occur abruptly but more gradually, and does not require nucleation to trigger it.

Sneutrino. The supersymmetric partner of the neutrino. Not yet observed.

Space Translation Symmetry. The laws of physics are the same at each point in space.

Special Relativity. Einstein's theory in which space, time, and mass are shown to depend on relative motion, and that energy and mass are equivalent, by $E = mc^2$.

Spectral Lines. The well-defined wavelengths at which molecules, atoms, and nuclei radiate, which are highly characteristic of the radiating substance.

Spin. An angular momentum in internal space that is a fundamental property of particles.

Spontaneously Broken Symmetry. A mechanism by which natural symmetries can be broken in a noncausal way.

Squark. The supersymmetric partner of the quark. Not yet observed.

Steady-State Universe. The theory that the universe is not expanding as in the Big Bang model, but the red shifts of galaxies are caused by other processes.

Strangeness. The quantum number carried by the *s* quark.

Strange Particle. Particle that has strangeness, such as kaon. Contains at least one *s* quark.

Strong Interaction or Strong Nuclear Force. The force between quarks that holds them together in particles and nuclei.

SU(2), SU(3), SU(5), etc. The names of mathematical symmetry groups associated with unification theories.

Superstrings. Fundamental objects that are not point particles but strings. A current fashionable candidate for the Theory of Everything.

Supersymmetry. The laws of physics are the same for fermions and bosons. An important ingredient in most schemes attempting to unify gravity with the other forces.

Time Dilation. The observed slowing of a moving clock in Einstein's Special Theory of Relativity.

Time Translation Symmetry. The law of physics are the same at all times.

Topological Defects. Places where the overall symmetry of a system is broken. Point defects are monopoles, line defects are strings, and planar defects are domain walls.

Tunneling. A quantum process by which a particle can pass through a barrier that is impossible in classical physics.

Unified Field Theory. Early attempts by Einstein and others to unify gravity and electromagnetism by geometrical means without quantum mechanics.

Wave Mechanics. Schrödinger's version of quantum mechanics in which the wave properties of particles are expressed by the wave function.

Weak Interaction or Weak Nuclear Force. Short-range interaction observed in nuclear and particle reactions, *e.g.,* beta decay. Now unified with electromagnetism.

Weak Isospin. An angular momentum in internal space that is associated with the symmetries of the weak interactions.

Weakly Interacting Massive Particles (WIMPs). Particles not yet observed that may constitute the major component of the matter of the universe.

References

These books and articles are either referred to directly in the text or were important sources of information or concepts to the author in writing this book.

Albrecht, A., and P. J. Steinhardt. *Physical Review Letters* 48 (1980):1220.

Angeles, P. A. *The Problem of God.* Prometheus Books (1980).

Arons, A. B., and A. M. Bork (eds.). *Science and Ideas.* Prentice-Hall (1964).

Asimov, Isaac. *Asimov's Guide to Halley's Comet.* Dell (1985).

Brown, L. M., and L. Hoddeson (eds.). *The Birth of Particle Physics.* Cambridge University Press (1983).

Cohen, I. B. (ed.). *Isaac Newton's Papers and Letters on Natural Philosophy.* Harvard University Press (1958).

Cooper, L. N. *An Introduction to the Meaning and Structure of Physics.* Harper and Row (1968).

Durant, W. *The Story of Philosophy.* Simon and Schuster (1953).

———. *The Story of Civilization.* Simon and Schuster (1944).

Ferris, T. *The Red Limit.* William Morrow (1977).

Frautschi, S. "Entropy in the Expanding Universe." *Science* 217 (1982): 593.

Freund, Philip. *Myths of Creation.* W. H. Allen (1964).

Guth, A. H. *Physical Review* D (1981): 23.

Hoyle, F., and C. Wickramasinghe. *Evolution from Space.* J. M. Dent and Sons (1981).

Kargon, R. H. *Atomism in England from Hariot to Newton.* Clarendon Press (1966).

Linde, A. D. *Physics Letters* B 108 (1982): 389.

Lucretius. *The Nature of Things.* Translation by F. O. Copley. W. W. Norton (1977).

Mason, S. F. *A History of the Sciences.* Routledge and Kegan Paul (1953).

Pais, A. *Subtle is the Lord.* Oxford (1982).

Segre, E. *From X-Rays to Quarks.* Freeman (1980).

Snyder, E. E. *A History of the Physical Sciences.* Merrill (1969).

Tryon, E. P. "Is the Universe a Vacuum Fluctuation?" *Nature* 246 (1973): 396.
Weinberg, S. *The First Three Minutes.* Basic Books (1977).

Suggested Reading

These books and articles were not major sources for the author, but present many of the issues discussed in the text in nontechnical language.

Calder, N. *The Key to the Universe.* Penguin (1977).

Davies, P. C. W. *The Accidental Universe.* Cambridge (1982).

———. "Relics of creation." *Sky and Telescope* 69, no. 2 (February 1985): 112.

———. "New Physics and the New Big Bang." *Sky and Telescope* 70, no. 5 (November 1985): 406.

Eddington, Arthur. *The Nature of the Physical World.* Cambridge (1928).

Freedman, D. Z., and P. van Nieuwenhuizen. "Supergravity and the Unification of the Laws of Physics." *Scientific American* 238, no. 2 (April 1978): 126.

———. "The Hidden Dimensions of Spacetime." *Scientific American* 252, no. 3 (March 1985): 74.

Georgi, H. "A Unified Theory of Elementary Particles and Forces." *Scientific American* 244, no. 4 (April 1981): 48.

Guth, A., and P. J. Steinhardt. "The Inflationary Universe." *Scientific American* 250, no. 5 (May 1984): 116.

McGowan, C. *In the Beginning . . .* Prometheus Books (1984).

Odenwald, S. "Does Space Have More Than 3 Dimensions?" *Astronomy* 12, no. 11 (November 1984): 66.

Pagels, H. *Perfect Symmetry.* Simon and Schuster (1985).

Ruse, M. *Darwinism Defended.* Addison-Wesley (1982).

Sciama, D. W. *The Physical Foundations of General Relativity.* Doubleday (1969).

Silk, J. *The Big Bang.* Freeman (1980).